Water Pollution: Effects, Control and Treatment

Water Pollution: Effects, Control and Treatment

Margaret Barton

Larsen & Keller
www.larsen-keller.com

Water Pollution: Effects, Control and Treatment
Margaret Barton
ISBN: 978-1-64172-421-0 (Hardback)

■ Larsen & Keller

Published by Larsen and Keller Education,
5 Penn Plaza,
19th Floor,
New York, NY 10001, USA

Cataloging-in-Publication Data

Water pollution : effects, control and treatment / Margaret Barton.
    p. cm.
Includes bibliographical references and index.
ISBN 978-1-64172-421-0
1. Water--Pollution. 2. Pollution prevention. 3. Water--Purification.
4. Water quality management. I. Barton, Margaret.
TD420 .W38 2020
363.739 4--dc23

For more information regarding Larsen and Keller Education and its products, please visit the publisher's website www.larsen-keller.com

Table of Contents

Preface

It is with great pleasure that I present this book. It has been carefully written after numerous discussions with my peers and other practitioners of the field. I would like to take this opportunity to thank my family and friends who have been extremely supporting at every step in my life.

The contamination of water bodies due to human activities is referred to as water pollution. It is majorly caused due to the discharge of wastewater into natural water bodies. The anthropogenic contaminants along with natural phenomena such as volcanoes, earthquakes, storms and algae blooms can cause changes in the quality and the ecological status of the water. Other contaminants include acidity, petroleum, detergents, food processing waste, ammonia, fertilizers and garbage. Chemicals and pathogens are also some pollutants that contribute to water pollution. Various physical, chemical and biological tests are used to analyze the water samples to measure the water pollution. This book provides comprehensive insights into the field of water pollution. Most of the topics introduced herein cover new techniques used for the control and treatment of water pollution. The book is appropriate for those seeking detailed information in this area.

The chapters below are organized to facilitate a comprehensive understanding of the subject:

Chapter – What is Water Pollution?

Water pollution can be defined as the contamination of water bodies such as rivers, lakes and oceans generally due to human activities. Major water pollutants include a wide range of pathogens and chemicals. This is an introductory chapter which will briefly introduce all the diverse aspects of water pollution.

Chapter – Types of Water Pollution

Some of the common types of water pollution are surface water pollution, ground water pollution, chemical water pollution, marine pollution, oil spill water pollution, microbial water pollution, etc. It occurs when harmful chemical particles or industrial, agricultural and residential waste enter the water bodies. This chapter has been carefully written to provide an easy understanding of these various types of water pollution.

Chapter – Sources of Water Pollution

The main sources of water pollution are domestic effluents and sewage, industrial effluents, agricultural effluents, radioactive wastes, thermal pollution and oil pollution. The topics elaborated in this chapter will help in gaining a better perspective about these sources of water pollution.

Chapter – Effects of Water Pollution

The effects of water pollution depend on the chemicals and concentration of water pollutants. Some of the adverse effects which can be caused by water pollution are killing of aquatic animals, disruption of food chains and their habitat, infectious diseases and destruction of ecosystems. All these effects of water pollution have been carefully analyzed in this chapter.

Chapter – Diseases Caused by Water Pollution

Water pollution can adversely affect the health of humans by causing numerous diseases. Some of the water-borne diseases are cholera, dysentery, diarrhea, typhoid, jaundice, amoebiasis, malaria, etc. This chapter closely examines all these diseases caused by water pollution to provide an extensive understanding of the subject.

Chapter – Prevention and Control of Water Pollution

Some of the ways that can be used to control and prevent water pollution are management of industrial waste, avoiding hazardous material and disposal of toxic wastes, cleaning of drains, recycling or reusing water. The aim of this chapter is to explore these various methods for prevention and control of water pollution.

Chapter – Water Treatment

The process used to improve the quality of water to make it potable for various uses is referred to as water treatment. The most common methods that are used for water treatment are sedimentation, filtration, aeration and chlorination. This chapter has been carefully written to provide an easy understanding of these methods of water treatment.

Margaret Barton

1

What is Water Pollution?

Water pollution can be defined as the contamination of water bodies such as rivers, lakes and oceans generally due to human activities. Major water pollutants include a wide range of pathogens and chemicals. This is an introductory chapter which will briefly introduce all the diverse aspects of water pollution.

Over two thirds of Earth's surface is covered by water; less than a third is taken up by land. As Earth's population continues to grow, people are putting ever-increasing pressure on the planet's water resources. In a sense, our oceans, rivers, and other inland waters are being "squeezed" by human activities—not so they take up less room, but so their quality is reduced. Poorer water quality means water pollution.

We know that pollution is a human problem because it is a relatively recent development in the planet's history: before the 19th century Industrial Revolution, people lived more in harmony with their immediate environment. As industrialization has spread around the globe, so the problem of pollution has spread with it. When Earth's population was much smaller, no one believed pollution would ever present a serious problem. It was once popularly believed that the oceans were far too big to pollute. Today, with around 7 billion people on the planet, it has become apparent that there are limits. Pollution is one of the signs that humans have exceeded those limits.

How serious is the problem? According to the environmental campaign organization WWF: "Pollution from toxic chemicals threatens life on this planet. Every ocean and every continent, from the tropics to the once-pristine polar regions, is contaminated.

Water pollution can be defined in many ways. Usually, it means one or more substances have built up in water to such an extent that they cause problems for animals or people. Oceans, lakes, rivers, and other inland waters can naturally clean up a certain amount of pollution by dispersing it harmlessly. If you poured a cup of black ink into a river, the ink would quickly disappear into the river's much larger volume of clean water. The ink would still be there in the river, but in such a low concentration that you would not be able to see it. At such low levels, the chemicals in the ink probably would not present any real problem. However, if you poured gallons of ink into a river every few seconds through a pipe, the river would quickly turn black. The chemicals in the ink could very quickly have an effect on the quality of the water. This, in turn, could affect the health of all the plants, animals, and humans whose lives depend on the river.

Thus, water pollution is all about *quantities*: how much of a polluting substance is released and how big a volume of water it is released into. A small quantity of a toxic chemical may have little impact if it is spilled into the ocean from a ship. But the same amount of the same chemical can have a much bigger impact pumped into a lake or river, where there is less clean water to disperse it.

Water pollution almost always means that some damage has been done to an ocean, river, lake, or other water source. A 1969 United Nations report defined ocean pollution as:

> "The introduction by man, directly or indirectly, of substances or energy into the marine environment (including estuaries) resulting in such deleterious effects as harm to living resources, hazards to human health, hindrance to marine activities, including fishing, impairment of quality for use of sea water and reduction of amenities."

Fortunately, Earth is forgiving and damage from water pollution is often reversible.

What are the main Types of Water Pollution?

When we think of Earth's water resources, we think of huge oceans, lakes, and rivers. Water resources like these are called surface waters. The most obvious type of water pollution affects surface waters. For example, a spill from an oil tanker creates an oil slick that can affect a vast area of the ocean.

Not all of Earth's water sits on its surface, however. A great deal of water is held in underground rock structures known as aquifers, which we cannot see and seldom think about. Water stored underground in aquifers is known as groundwater. Aquifers feed our rivers and supply much of our drinking water. They too can become polluted, for example, when weed killers used in people's gardens drain into the ground. Groundwater pollution is much less obvious than surface-water pollution, but is no less of a problem. In 1996, a study in Iowa in the United

States found that over half the state's groundwater wells were contaminated with weed killers. You might think things would have improved since then, but, two decades on, all that's really changed is the name of the chemicals we're using. Today, numerous scientific studies are still finding weed killers in groundwater in worrying quantities: a 2012 study discovered glyphosate in 41 percent of 140 groundwater samples from Catalonia, Spain; scientific opinion differs on whether this is safe or not.

Surface waters and groundwater are the two types of water resources that pollution affects. There are also two different ways in which pollution can occur. If pollution comes from a single location, such as a discharge pipe attached to a factory, it is known as point-source pollution. Other examples of point source pollution include an oil spill from a tanker, a discharge from a smoke stack (factory chimney), or someone pouring oil from their car down a drain. A great deal of water pollution happens not from one single source but from many different scattered sources. This is called nonpoint-source pollution.

Above: Point-source pollution comes from a single, well-defined place such as this pipe.
Below: Nonpoint-source pollution comes from many sources. All the industrial plants alongside a river and the ships that service them may be polluting the river collectively.

When point-source pollution enters the environment, the place most affected is usually the area immediately around the source. For example, when a tanker accident occurs, the oil slick is concentrated around the tanker itself and, in the right ocean conditions, the pollution disperses the further away from the tanker you go. This is less likely to happen with nonpoint source pollution which, by definition, enters the environment from many different places at once.

Sometimes pollution that enters the environment in one place has an effect hundreds or even thousands of miles away. This is known as transboundary pollution. One example is the way radioactive waste travels through the oceans from nuclear reprocessing plants in England and France to nearby countries such as Ireland and Norway.

How do we know when Water is Polluted?

Some forms of water pollution are very obvious: everyone has seen TV news footage of oil slicks filmed from helicopters flying overhead. Water pollution is usually less obvious and much harder to detect than this. But how can we measure water pollution when we cannot see it? How do we even know it's there?

There are two main ways of measuring the quality of water. One is to take samples of the water and measure the concentrations of different chemicals that it contains. If the chemicals are dangerous or the concentrations are too great, we can regard the water as polluted. Measurements like this are known as chemical indicators of water quality. Another way to measure water quality involves examining the fish, insects, and other invertebrates that the water will support. If many different types of creatures can live in a river, the quality is likely to be very good; if the river supports no fish life at all, the quality is obviously much poorer. Measurements like this are called biological indicators of water quality.

What are the causes of Water Pollution?

Most water pollution doesn't begin in the water itself. Take the oceans: around 80 percent of ocean pollution enters our seas from the land. Virtually any human activity can have an effect on the quality of our water environment. When farmers fertilize the fields, the chemicals they use are gradually washed by rain into the groundwater or surface waters nearby. Sometimes the causes of water pollution are quite surprising. Chemicals released by smokestacks (chimneys) can enter the atmosphere and then fall back to earth as rain, entering seas, rivers, and lakes and causing water pollution. That›s called atmospheric deposition. Water pollution has many different causes and this is one of the reasons why it is such a difficult problem to solve.

Sewage

With billions of people on the planet, disposing of sewage waste is a major problem. According to 2015 figures from the World Health Organization (the most recent available at the time this article was updated in 2019), some 2.1 billion people (28 percent of the world's population) don't have access to safe drinking water, while 2.3 billion (30 percent of the world's population) don't have proper sanitation (hygienic toilet facilities); although there have been great improvements in securing access to clean water, relatively little progress has been made on improving global sanitation in the last decade. Sewage disposal affects people›s immediate environments and leads to water-related illnesses such as diarrhea that kills 525,000 children under five each year. (Back in 2002, the World Health Organization estimated that water-related diseases could kill as many as 135 million people by 2020; in 2016, the WHO was still estimating the annual death toll from poor water and sanitation at close to a million people a year.) In developed countries, most people have flush toilets that take sewage waste quickly and hygienically away from their homes.

Yet the problem of sewage disposal does not end there. When you flush the toilet, the waste has to go somewhere and, even after it leaves the sewage treatment works, there is still waste to dispose of. Sometimes sewage waste is pumped untreated into the sea. Until the early 1990s, around 5 million tons of sewage was dumped by barge from New York City each year. According to 2002 figures from the UK government›s Department for the Environment, Food, and Rural Affairs (DEFRA), the sewers of Britain collect around 11 billion liters of waste water every day; there are still 31,000 sewage overflow pipes through which, in certain circumstances, raw sewage is pumped untreated into the sea. The New River that crosses the border from Mexico into California once carried with it 20–25 million gallons (76–95 million liters) of raw sewage each day; a new waste water plant on the US-Mexico border, completed in 2007, substantially solved that problem. Unfortunately, even in some of the richest nations, the practice of dumping sewage into the sea continues. In early 2012, it was reported that the tiny island of Guernsey (between Britain and France) has decided to continue dumping 16,000 tons of raw sewage into the sea each day.

In theory, sewage is a completely natural substance that should be broken down harmlessly in the environment: 90 percent of sewage is water. In practice, sewage contains all kinds of other chemicals, from the pharmaceutical drugs people take to the paper, plastic, and other wastes they flush down their toilets. When people are sick with viruses, the sewage they produce carries those viruses into the environment. It is possible to catch illnesses such as hepatitis, typhoid, and cholera from river and sea water.

Nutrients

During crop-spraying, some chemicals will drain into the soil. Eventually, they seep into rivers and other watercourses.

Suitably treated and used in moderate quantities, sewage can be a fertilizer: it returns important nutrients to the environment, such as nitrogen and phosphorus, which plants and animals need for growth. The trouble is, sewage is often released in much greater quantities than the natural environment can cope with. Chemical fertilizers used by farmers also add nutrients to the soil, which drain into rivers and seas and add to the fertilizing effect of the sewage. Together, sewage and fertilizers can cause a massive increase in the growth of algae or plankton that overwhelms huge areas of oceans, lakes, or rivers. This is known as a harmful algal bloom (also known as an HAB or red tide, because it can turn the water red). It is harmful because it removes oxygen from the water that kills other forms of life, leading to what is known as a dead zone. The Gulf of Mexico has one of the world's most spectacular dead zones. Each summer, according to studies by the NOAA, it grows to an area of around 5500–6000 square miles (14,000–15,500 square kilometers), which is about the same size as the state of Connecticut.

Waste Water

A few statistics illustrate the scale of the problem that waste water (chemicals washed down drains and discharged from factories) can cause. Around half of all ocean pollution is caused by sewage and waste water. Each year, the world generates perhaps 5–10 billion tons of industrial waste, much of which is pumped untreated into rivers, oceans, and other waterways. In the United States alone, around 400,000 factories take clean water from rivers, and many pump polluted waters back in their place. However, there have been major improvements in waste water treatment recently. Since 1970, in the United States, the Environmental Protection Agency (EPA) has invested about $70 billion in improving water treatment plants that, as of 2015, serve around 88 percent of the US population (compared to just 69 percent in 1972). However, another $271 billion is still needed to update and upgrade the system.

Factories are point sources of water pollution, but quite a lot of water is polluted by ordinary people from nonpoint sources; this is how ordinary water becomes waste water in the first place. Virtually everyone pours chemicals of one sort or another down their drains or toilets. Even detergents used in washing machines and dishwashers eventually end up in our rivers and oceans. So do the pesticides we use on our gardens. A lot of toxic pollution also enters waste water from highway runoff. Highways are typically covered with a cocktail of toxic chemicals—everything from spilled fuel and brake fluids to bits of worn tires (themselves made from chemical additives) and exhaust emissions. When it rains, these chemicals wash into drains and rivers. It is not unusual for heavy summer rainstorms to wash toxic chemicals into rivers in such concentrations that they kill large numbers of fish overnight. It has been estimated that, in one year, the highway runoff from a single large city leaks as much oil into our water environment as a typical tanker spill. Some highway runoff runs away into drains; others can pollute groundwater or accumulate in the land next to a road, making it increasingly toxic as the years go by.

Chemical Waste

Detergents are relatively mild substances. At the opposite end of the spectrum are highly toxic chemicals such as polychlorinated biphenyls (PCBs). They were once widely used to manufacture electronic circuit boards, but their harmful effects have now been recognized and their use is highly restricted in many countries. Nevertheless, an estimated half million tons of PCBs were discharged into the environment during the 20th century. In a classic example of transboundary pollution, traces of PCBs have even been found in birds and fish in the Arctic. They were carried there through the oceans, thousands of miles from where they originally entered the environment. Although PCBs are widely banned, their effects will be felt for many decades because they last a long time in the environment without breaking down.

Another kind of toxic pollution comes from heavy metals, such as lead, cadmium, and mercury. Lead was once commonly used in gasoline (petrol), though its use is now restricted in some countries. Mercury and cadmium are still used in batteries (though some brands now use other metals instead). Until recently, a highly toxic chemical called tributyltin (TBT) was used in paints to protect boats from the ravaging effects of the oceans. Ironically, however, TBT was gradually recognized as a pollutant: boats painted with it were doing as much damage to the oceans as the oceans were doing to the boats.

The best known example of heavy metal pollution in the oceans took place in 1938 when a Japanese factory discharged a significant amount of mercury metal into Minamata Bay, contaminating the fish stocks there. It took a decade for the problem to come to light. By that time, many local people had eaten the fish and around 2000 were poisoned. Hundreds of people were left dead or disabled.

Radioactive Waste

People view radioactive waste with great alarm—and for good reason. At high enough concentrations it can kill; in lower concentrations it can cause cancers and other illnesses. The biggest sources of radioactive pollution in Europe are two factories that reprocess waste fuel from nuclear power plants: Sellafield on the north-west coast of Britain and Cap La Hague on the north coast of France. Both discharge radioactive waste water into the sea, which ocean currents then carry around the world. Countries such as Norway, which lie downstream from Britain, receive significant doses of radioactive pollution from Sellafield. The Norwegian government has repeatedly complained that Sellafield has increased radiation levels along its coast by 6–10 times. Both the Irish and Norwegian governments continue to press for the plant's closure.

Oil Pollution

Oil-tanker spills are the most spectacular forms of pollution and the ones
that catch public attention, but only a fraction of all water pollution happens this way.

When we think of ocean pollution, huge black oil slicks often spring to mind, yet these spectacular accidents represent only a tiny fraction of all the pollution entering our oceans. Even considering oil by itself, tanker spills are not as significant as they might seem: only 12 percent of the oil that enters the oceans comes from tanker accidents; over 70 percent of oil pollution at sea comes from routine shipping and from the oil people pour down drains on land. However, what makes tanker spills so destructive is the sheer quantity of oil they release *at once* — in other words, the concentration of oil they produce in one very localized part of the marine environment. The biggest oil spill in recent years (and the biggest ever spill in US waters) occurred when the tanker *Exxon Valdez* broke up in Prince William Sound in Alaska in 1989. Around 12 million gallons (44 million liters) of oil were released into the pristine wilderness—enough to fill your living room 800 times over! Estimates of the marine animals killed in the spill vary from approximately 1000 sea otters and 34,000 birds to as many as 2800 sea otters and 250,000 sea birds. Several billion salmon and herring eggs are also believed to have been destroyed.

Plastics

If you've ever taken part in a community beach clean, you'll know that plastic is far and away the most common substance that washes up with the waves. There are three reasons for this: plastic is one of the most common materials, used for making virtually every kind of manufactured object from clothing to automobile parts; plastic is light and floats easily so it can travel enormous distances across the oceans; most plastics are not biodegradable (they do not break down naturally in the environment), which means that things like plastic bottle tops can survive in the marine environment for a long time. (A plastic bottle can survive an estimated 450 years in the ocean and plastic fishing line can last up to 600 years.)

While plastics are not toxic in quite the same way as poisonous chemicals, they nevertheless present a major hazard to seabirds, fish, and other marine creatures. For example, plastic fishing lines and other debris can strangle or choke fish. (This is sometimes called ghost fishing.) About half of all the world's seabird species are known to have eaten plastic residues. In one study of 450 shear-waters in the North Pacific, over 80 percent of the birds were found to contain plastic residues in their stomachs. In the early 1990s, marine scientist Tim Benton collected debris from a 2km (1.5 mile) length of beach in the remote Pitcairn islands in the South Pacific. His study recorded approximately a thousand pieces of garbage including 268 pieces of plastic, 71 plastic bottles, and two dolls heads.

Today, much media attention focuses on the Great Pacific Garbage Patch, a floating, oceanic graveyard of plastic junk roughly three times the size of France, discovered by sailor Charles J. Moore in 1997. But, as you›ll know well enough if you›ve ever taken part in a community beach clean, persistent plastic litters every ocean on the planet: some 8 million tons of new plastic are dumped in the sea every single year.

Alien Species

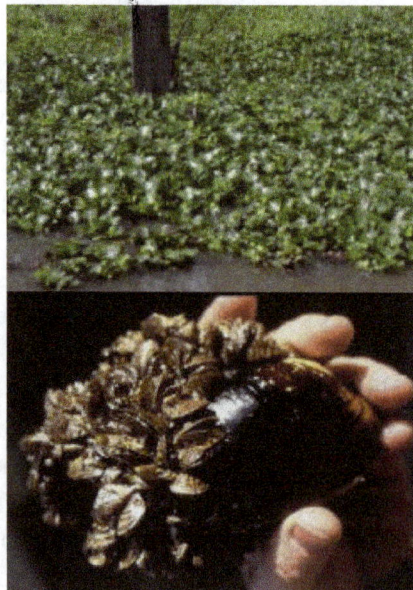

Invasive species: Above: Water hyacinth crowding out a waterway around an old fence post.
Below: Non-native zebra mussels clumped on a native mussel.

Most people's idea of water pollution involves things like sewage, toxic metals, or oil slicks, but pollution can be biological as well as chemical. In some parts of the world, alien species are a major problem. Alien species (sometimes known as invasive species) are animals or plants from one region that have been introduced into a different ecosystem where they do not belong. Outside their normal environment, they have no natural predators, so they rapidly run wild, crowding out the usual animals or plants that thrive there. Common examples of alien species include zebra mussels in the Great Lakes of the USA, which were carried there from Europe by ballast water (waste water flushed from ships). The Mediterranean Sea has been invaded by a kind of alien algae called Caulerpa taxifolia. In the Black Sea, an alien jellyfish called Mnemiopsis leidyi reduced fish stocks by 90 percent after arriving in ballast water. In San Francisco Bay, Asian clams called *Potamocorbula amurensis*, also introduced by ballast water, have dramatically altered the ecosystem. In 1999, Cornell University›s David Pimentel estimated that alien invaders like this cost the US economy $123 billion a year; in 2014, the European Commission put the cost to Europe at €12 billion a year and «growing all the time.

Other Forms of Pollution

These are the most common forms of pollution—but by no means the only ones. Heat or thermal pollution from factories and power plants also causes problems in rivers. By raising the temperature, it reduces the amount of oxygen dissolved in the water, thus also reducing the level of aquatic life that the river can support.

Another type of pollution involves the disruption of sediments (fine-grained powders) that flow from rivers into the sea. Dams built for hydroelectric power or water reservoirs can reduce the sediment flow. This reduces the formation of beaches, increases coastal erosion (the natural destruction of cliffs by the sea), and reduces the flow of nutrients from rivers into seas (potentially reducing coastal fish stocks). Increased sediments can also present a problem. During construction work, soil, rock, and other fine powders sometimes enters nearby rivers in large quantities, causing it to become turbid (muddy or silted). The extra sediment can block the gills of fish, effectively suffocating them. Construction firms often now take precautions to prevent this kind of pollution from happening.

What are the Effects of Water Pollution?

Some people believe pollution is an inescapable result of human activity: they argue that if we want to have factories, cities, ships, cars, oil, and coastal resorts, some degree of pollution is almost certain to result. In other words, pollution is a necessary evil that people must put up with if they want to make progress. Fortunately, not everyone agrees with this view. One reason people have woken up to the problem of pollution is that it brings costs of its own that undermine any economic benefits that come about by polluting.

Take oil spills, for example. They can happen if tankers are too poorly built to survive accidents at sea. But the economic benefit of compromising on tanker quality brings an economic cost when an oil spill occurs. The oil can wash up on nearby beaches, devastate the ecosystem, and severely affect tourism. The main problem is that the people who bear the cost of the spill (typically a small coastal community) are not the people who caused the problem in the first place (the people who operate the tanker). Yet, arguably, everyone who puts gasoline (petrol) into their car—or uses almost any kind of petroleum-fueled transport—contributes to the problem in some way. So oil spills are a problem for everyone, not just people who live by the coast and tanker operates.

Sewage is another good example of how pollution can affect us all. Sewage discharged into coastal waters can wash up on beaches and cause a health hazard. People who bathe or surf in the water can fall ill if they swallow polluted water—yet sewage can have other harmful effects too: it can poison shellfish (such as cockles and mussels) that grow near the shore. People who eat poisoned shellfish risk suffering from an acute—and sometimes fatal—illness called paralytic shellfish poisoning. Shellfish is no longer caught along many shores because it is simply too polluted with sewage or toxic chemical wastes that have discharged from the land nearby.

Pollution matters because it harms the environment on which people depend. The environment is not something distant and separate from our lives. It's not a pretty shoreline hundreds of miles from our homes or a wilderness landscape that we see only on TV. The environment is everything that surrounds us that gives us life and health. Destroying the environment ultimately reduces the quality of our own lives—and that, most selfishly, is why pollution should matter to all of us.

How can we Stop Water Pollution?

There is no easy way to solve water pollution; if there were, it wouldn't be so much of a problem. Broadly speaking, there are three different things that can help to tackle the problem—education, laws, and economics—and they work together as a team.

Education

Making people aware of the problem is the first step to solving it. In the early 1990s, when surfers in Britain grew tired of catching illnesses from water polluted with sewage, they formed a group called Surfers Against Sewage to force governments and water companies to clean up their act. People who've grown tired of walking the world's polluted beaches often band together to organize community beach-cleaning sessions. Anglers who no longer catch so many fish have campaigned for tougher penalties against factories that pour pollution into our rivers. Greater public awareness can make a positive difference.

Laws

One of the biggest problems with water pollution is its transboundary nature. Many rivers cross countries, while seas span whole continents. Pollution discharged by factories in one country with poor environmental standards can cause problems in neighboring nations, even when they have tougher laws and higher standards. Environmental laws can make it tougher for people to pollute, but to be really effective they have to operate across national and international borders. This is why we have international laws governing the oceans, such as the 1982 UN Convention on the Law of the Sea (signed by over 120 nations), the 1972 London (Dumping) Convention, the 1978 MARPOL International Convention for the Prevention of Pollution from Ships, and the 1998 OSPAR Convention for the Protection of the Marine Environment of the North East Atlantic. The European Union has water-protection laws (known as directives) that apply to all of its member states. They include the 1976 Bathing Water Directive (updated 2006), which seeks to ensure the quality of the waters that people use for recreation. Most countries also have their own water pollution laws. In the United States, for example, there is the 1972 Clean Water Act and the 1974 Safe Drinking Water Act.

Economics

Most environmental experts agree that the best way to tackle pollution is through something called the polluter pays principle. This means that whoever causes pollution should have to pay to clean it up, one way or another. Polluter pays can operate in all kinds of ways. It could mean that tanker owners should have to take out insurance that covers the cost of oil spill cleanups, for example. It could also mean that shoppers should have to pay for their plastic grocery bags, as is now common in Ireland, to encourage recycling and minimize waste. Or it could mean that factories that use rivers must have their water inlet pipes downstream of their effluent outflow pipes, so if they cause pollution they themselves are the first people to suffer. Ultimately, the polluter pays principle is designed to deter people from polluting by making it less expensive for them to behave in an environmentally responsible way.

Our Clean Future

Life is ultimately about choices—and so is pollution. We can live with sewage-strewn beaches, dead rivers, and fish that are too poisonous to eat. Or we can work together to keep the environment clean so the plants, animals, and people who depend on it remain healthy. We can take individual action to help reduce water pollution, for example, by using environmentally friendly detergents, not pouring oil down drains, reducing pesticides, and so on. We can take community action too, by helping out on beach cleans or litter picks to keep our rivers and seas that little bit cleaner. And we can take action as countries and continents to pass laws that will make pollution harder and the world less polluted. Working together, we can make pollution less of a problem—and the world a better place.

Types of Water Pollutants

The various types of water pollutants can be classified in to following major categories: (1) Organic pollutants, (2) Pathogens, (3) Nutrients and agriculture runoff, (4) Suspended solids and sediments (organic and inorganic), (5) Inorganic pollutants (salts and metals), (6) Thermal Pollution, and (7) Radioactive pollutants.

Organic Pollutants

Organic pollutants can be further divided into following categories:

a) Oxygen Demanding wastes: The wastewaters such as, domestic and municipal sewage, waste-water from food processing industries, canning industries, slaughter houses, paper and pulp mills, tanneries, breweries, distilleries, etc. have considerable concentration of biodegradable organic compounds either in suspended, colloidal or dissolved form. These wastes undergo degradation and decomposition by bacterial activity.

The dissolved oxygen available in the water body will be consumed for aerobic oxidation of organic matter present in the wastewater. Hence, depletion of the DO will be a serious problem adversely affecting aquatic life, if the DO falls below 4.0 mg/L. This decrease of DO is an index of pollution.

b) Synthetic Organic Compounds: Synthetic organic compounds are also likely to enter the ecosystem through various manmade activities such as production of these compounds, spillage during transportation, and their uses in different applications. These include synthetic pesticides, synthetic detergents, food additives, pharmaceuticals, insecticides, paints, synthetic fibers, plastics, solvents and volatile organic compounds (VOCs). Most of these compounds are toxic and biorefractory organics i.e., they are resistant to microbial degradation.

Even concentration of some of these in traces may make water unfit for different uses. The detergents can form foams and volatile substances may cause explosion in sewers. Polychlorinated biphenyls (PCBs) are used in the industries since 1930s which are complex mixtures of chlorobiphenyls. Being a fat soluble they move readily through the environment and within the tissues or cells. Once introduced into environment, these compounds are exceedingly persistent and their stability to chemical reagents is also high.

c) Oil: Oil is a natural product which results from the plant remains fossilized over millions of years, under marine conditions. It is a complex mixture of hydrocarbons and degradable under bacterial action, the biodegradation rate is different for different oils, tars being one of the slowest. Oil enters in to water through oil spills, leak from oil pipes, and wastewater from production and refineries.

Being lighter than water it spreads over the surface of water, separating the contact of water with air, hence resulting in reduction of DO. This pollutant is also responsible for endangering water birds and coastal plants due to coating of oils and adversely affecting the normal activities. It also results in reduction of light transmission through surface waters, thereby reducing the photosynthetic activity of the aquatic plants. Oil includes polycyclic aromatic hydrocarbons (PAH), some of which are known to be carcinogenic.

Pathogens

The pathogenic microorganisms enter in to water body through sewage discharge as a major source or through the wastewater from industries like slaughterhouses. Viruses and bacteria can cause water borne diseases, such as cholera, typhoid, dysentery, polio and infectious hepatitis in human.

Nutrients

The agriculture run-off, wastewater from fertilizer industry and sewage contains substantial concentration of nutrients like nitrogen and phosphorous. These waters supply nutrients to the plants and may stimulate the growth of algae and other aquatic weeds in receiving waters. Thus, the value of the water body is degraded. In long run, water body reduces DO, leads to eutrophication and ends up as a dead pool of water. People swimming in eutrophic waters containing blue-green algae can have skin and eye irritation, gastroenteritis and vomiting. High nitrogen levels in the water supply, causes a potential risk, especially to infants under six months. This is when the methemoglobin results in a decrease in the oxygen carrying capacity of the blood (blue baby disease) as nitrate ions in the blood readily oxidize ferrous ions in the hemoglobin.

In freshwater systems, eutrophication is a process whereby water bodies receive excess inorganic nutrients, especially N and P, which stimulate excessive growth of plants and algae. Eutrophication can happen naturally in the normal succession of some freshwater ecosystems. However, when the nutrient enrichment is due to the activities of humans, sometimes referred to as "cultural eutrophication", the rate of this natural process is greatly intensified. Two major nutrients, nitrogen (N) and phosphorus (P), occur in streams in various forms as ions or dissolved in solution. Aquatic plants convert dissolved inorganic forms of nitrogen (nitrate, nitrite, and ammonium) and phosphorus (orthophosphate) into organic or particulate forms for use in higher trophic production. The main effects caused by eutrophication can be summarized as follows:

- Species diversity decreases and the dominant biota changes.

- Plant and animal biomass increase.

- Turbidity increases.

- Rate of sedimentation increases, shortening the lifespan of the lake.

- Anoxic conditions may develop.

Suspended Solids and Sediments

These comprise of silt, sand and minerals eroded from land. These appear in the water through the surface runoff during rainy season and through municipal sewers. This can lead to the siltation, reduces storage capacities of reservoirs. Presence of suspended solids can block the sunlight penetration in the water, which is required for the photosynthesis by bottom vegetation. Deposition of the solids in the quiescent stretches of the stream or ocean bottom can impair the normal aquatic life and affect the diversity of the aquatic ecosystem. If the deposited solids are organic in nature, they will undergo decomposition leading to development of anaerobic conditions. Finer suspended solids such as silt and coal dust may injure the gills of fishes and cause asphyxiation.

Inorganic Pollutants

Apart from the organic matter discharged in the water body through sewage and industrial wastes, high concentration of heavy metals and other inorganic pollutants contaminate the water. These compounds are non-biodegradable and persist in the environment. These pollutants include mineral acids, inorganic salts, trace elements, metals, metals compounds, complexes of metals with organic compounds, cyanides, sulphates, etc.

- The accumulation of heavy metals may have adverse effect on aquatic flora and fauna and may constitute a public health problem where contaminated organisms are used for food.

- Algal growth due to nitrogen and phosphorous compounds can be observed.

- Metals in high concentration can be toxic to biota e.g. Hg, Cu, Cd, Pb, As, and Se.

- Copper greater than 0.1 mg/L is toxic to microbes.

Thermal Pollution

Considerable thermal pollution results due to discharge of hot water from thermal power plants, nuclear power plants, and industries where water is used as coolant. As a result of hot water discharge, the temperature of water body increases. Rise in temperature reduces the DO content of the water, affecting adversely the aquatic life. This alters the spectrum of organisms, which can adopt to live at that temperature and DO level. When organic matter is also present, the bacterial action increases due to rise in temperature; hence, resulting in rapid decrease of DO. The discharge of hot water leads to the thermal stratification in the water body, where hot water will remain on the top.

Radioactive Pollutants

Radioactive materials originate from the following:

- Mining and processing of ores.

- Use in research, agriculture, medical and industrial activities, such as I131, P32, Co60, Ca45, S35, C14, etc.

- Radioactive discharge from nuclear power plants and nuclear reactors, e.g., Sr90, Cesium Cs137, Plutonium Pu248, Uranium-238, Uranium-235.

- Uses and testing of nuclear weapons.

Various pollutants and their adverse effect:

Sr. No.	Pollutants	Impact
1.	Organic pollutants i)Oxygen Demanding wastes: ii)Synthetic organic pollutants iii) oil	Depletion of the DO will be a serious problem adversely affecting aquatic life, if the DO falls below 4.0 mg/L. Most of these compounds are toxic and biorefractory organics. It also make water unfit for different uses. This pollutant is also responsible for endangering water birds and coastal plants due to coating of oils and adversely affecting the normal activities which cause reduction of light transmission and photosynthesis.
2.	Pathogens	Number of diseases transmitted by pathogens available in wastewater
3.	Nutrients	When these are disposed in aquatic environment, it can lead to growth of undesirable aquatic life. When it discharged on land it causes groundwater pollution.
4.	Thermal pollutants	When organic matter is also present, the bacterial action increases due to rise in temperature; hence, resulting in rapid decrease of DO. It also results in thermal stratification which alters spectrum of organisms.
5.	Radioactive pollutants	These isotopes are toxic to the life forms; they accumulate in the bones, teeth and can cause serious disorders
6.	Suspended solids and sediments	Presence of suspended solids can block the sunlight penetration in the water, which is required for the photosynthesis by bottom vegetation. Finer suspended solids such as silt and coal dust may injure the gills of fishes and cause asphyxiation.
7.	Inorganic pollutants	These pollutants include mineral acids, inorganic salts, trace elements, metals, metals compounds, complexes of metals with organic compounds, cyanides, sulphates, etc. They have adverse effect on aquatic flora and fauna and may constitute a public health problem.

These isotopes are toxic to the life forms; they accumulate in the bones, teeth and can cause serious disorders. The safe concentration for lifetime consumption is 1 x 10-7 microcuries per ml. The summary of various pollutants and their adverse effect on the environment is presented in table. The major impacts have been described, however there are additional adverse effects of release of these pollutants in the environment.

2
Types of Water Pollution

Some of the common types of water pollution are surface water pollution, ground water pollution, chemical water pollution, marine pollution, oil spill water pollution, microbial water pollution, etc. It occurs when harmful chemical particles or industrial, agricultural and residential waste enter the water bodies. This chapter has been carefully written to provide an easy understanding of these various types of water pollution.

The different types of water pollution can be broadly classified as point-source, non-point source and, trans-boundary. In simple terms, it means that, if the water is polluted from a single point, such as an oil spill, then it is termed as a point-source based water pollution. Whereas, if there are multiple sources, surrounding an area from which pollutants are entering the water, then it is non-point-source water pollution. Trans-boundary pollution is when the water gets polluted even many miles away from where the pollutant source is, as in the case of nuclear waste.

Surface Water Pollution

Surface water contamination is a significant problem for municipal authorities. It can lead to pollution on a large scale and be a cause of major illness in humans. Knowing how to recognize it, minimize its occurrence, and treat the problem effectively is vital to maintaining a safe water supply.

Water from lakes and rivers that are used by municipalities, agriculture, and industry, is increasingly exposed to pollutants from manufacturing or the environment.

Fertilizers can leak into rivers, and flooding leads to pollution of surface water as the volume spreads across areas that are normally not exposed to water. These contaminants are why water must be treated before being used for human consumption.

Even though water appears to be clear, it may not be clean, which is why municipal authorities need a program of testing and treatment for potential contamination.

Sources of Surface Water Pollution

How can surface water become contaminated? One of the most common sources of surface water pollution is human waste, especially in developing countries.

In addition to human waste, there are issues with fertilizer seepage from farmland into ground-water. Industrial plants are also known to contaminate surface water with byproducts leaking into rivers and drainage systems.

Poorly maintained waste systems and adverse weather incidents such as flooding are also major sources of surface water pollution. For a municipal authority, knowing the risks in the area and what to do about it is critical.

The sources of surface water contamination are many. Having municipal surface water treatment services available is vital to community health, as treatment specialists have the expertise to analyze and treat the problem efficiently and effectively.

Pathogens in Surface Water Pollution

One of the biggest risks to humans from surface water pollution are pathogens that cause types of waterborne diseases. These come from human waste, as well as industrial sources which include organic chemicals and heavy metals.

Contamination most commonly occurs when food is prepared using contaminated water or by a person drinking it. This is a common cause of illness, particularly in developing countries.

Surface water contamination can also lead to toxic products remaining in fish because of exposure to pathogens. Municipal water suppliers need to access the expertise of surface water treatment services to diagnose and treat the problem before it becomes a major hazard to health.

Color and Turbidity

The color and turbidity of the water are affected when there is contamination. Typically, tannins are formed from organic material and include branches, soil, fish, debris and more.

The type of tannin will depend on the location and nature of the contamination. Some are difficult to see so having regular checks of the water supply is important.

Turbidity occurs when there are sediments in the water which typically settle or cloud the appearance of the liquid. This is not necessarily harmful. Turbidity is more common in surface water as it lacks the natural filtration found in groundwater. Working with a municipal water treatment expert like AOS to address the problem will improve the quality of the supply.

A municipal authority should use the expertise of water treatment services to maintain a safe water supply and reduce the risk of contamination. Surface water is exposed to several contaminants, depending on the location so regular inspections and treatments will improve quality

Ground Water Pollution

Groundwater pollution (also called groundwater contamination) occurs when pollutants are released to the ground and make their way down into groundwater. This type of water pollution can also occur naturally due to the presence of a minor and unwanted constituent, contaminant or impurity in the groundwater, in which case it is more likely referred to as contamination rather than pollution.

The pollutant often creates a contaminant plume within an aquifer. Movement of water and dispersion within the aquifer spreads the pollutant over a wider area. Its advancing boundary, often called a plume edge, can intersect with groundwater wells or daylight into surface water such as seeps and springs, making the water supplies unsafe for humans and wildlife. The movement of the plume, called a plume front, may be analyzed through a hydrological transport model or groundwater model. Analysis of groundwater pollution may focus on soil characteristics and site geology, hydrogeology, hydrology, and the nature of the contaminants.

Pollution can occur from on-site sanitation systems, landfills, effluent from wastewater treatment plants, leaking sewers, petrol filling stations or from over application of fertilizers in agriculture. Pollution (or contamination) can also occur from naturally occurring contaminants, such as arsenic or fluoride. Using polluted groundwater causes hazards to public health through poisoning or the spread of disease.

Different mechanisms have influence on the transport of pollutants, e.g. diffusion, adsorption, precipitation, decay, in the groundwater. The interaction of groundwater contamination with surface waters is analyzed by use of hydrology transport models.

Pollutant Types

Contaminants found in groundwater cover a broad range of physical, inorganic chemical, organic chemical, bacteriological, and radioactive parameters. Principally, many of the same pollutants that play a role in surface water pollution may also be found in polluted groundwater, although their respective importance may differ.

Arsenic and Fluoride

Arsenic and fluoride have been recognized by the World Health Organization (WHO) as the most serious inorganic contaminants in drinking-water on a worldwide basis.

The metalloid arsenic can occur naturally in groundwater, as seen most frequently in Asia, including in China, India and Bangladesh. In the Ganges Plain of northern India and Bangladesh severe contamination of groundwater by naturally occurring arsenic affects 25% of water wells in the shallower of two regional aquifers.

Arsenic in groundwater can also be present where there are mining operations or mine waste dumps that will leach arsenic.

Natural fluoride in groundwater is of growing concern as deeper groundwater is being used, "with more than 200 million people at risk of drinking water with elevated concentrations." Fluoride can especially be released from acidic volcanic rocks and dispersed volcanic ash when water hardness is low. High levels of fluoride in groundwater is a serious problem in the Argentinean Pampas, Chile, Mexico, India, Pakistan, the East African Rift, and some volcanic islands (Tenerife).

In areas that have naturally occurring high levels of fluoride in groundwater which is used for drinking water, both dental and skeletal fluorosis can be prevalent and severe.

Pathogens

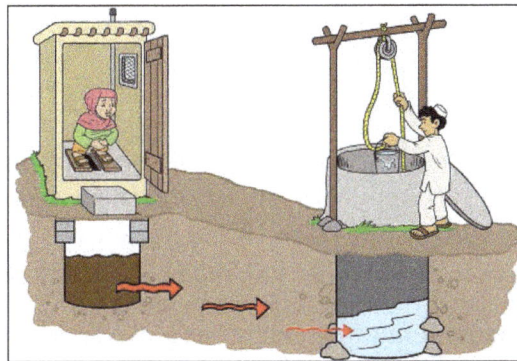

Waterborne diseases can be spread via a groundwater well which is contaminated with fecal pathogens from pit latrines.

The lack of proper sanitation measures, as well as improperly placed wells, can lead to drinking water contaminated with pathogens carried in feces and urine. Such fecal-oral transmitted diseases include cholera and diarrhea. Of the four pathogen types that are present in feces (bacteria, viruses, protozoa, and helminths or helminth eggs), the first three can be commonly found in polluted groundwater, whereas the relatively large helminth eggs are usually filtered out by the soil matrix.

Deep, confined aquifers are usually considered the safest source of drinking water with respect to pathogens. Pathogens from treated or untreated wastewater can contaminate certain, especially shallow, aquifers.

Nitrate

Nitrate is the most common chemical contaminant in the world's groundwater and aquifers. In some low-income countries nitrate levels in groundwater are extremely high, causing significant health problems. It is also stable (it does not degrade) under high oxygen conditions.

Nitrate levels above 10 mg/L (10 ppm) in groundwater can cause "blue baby syndrome" (acquired methemoglobinemia). Drinking water quality standards in the European Union stipulate less than 50 mg/L for nitrate in drinking water.

However, the linkages between nitrates in drinking water and blue baby syndrome have been disputed in other studies. The syndrome outbreaks might be due to other factors than elevated nitrate concentrations in drinking water.

Elevated nitrate levels in groundwater can be caused by on-site sanitation, sewage sludge disposal and agricultural activities. It can therefore have an urban or agricultural origin.

Organic Compounds

Volatile organic compounds (VOCs) are a dangerous contaminant of groundwater. They are generally introduced to the environment through careless industrial practices. Many of these compounds were not known to be harmful until the late 1960s and it was some time before regular testing of groundwater identified these substances in drinking water sources.

Primary VOC pollutants found in groundwater include aromatic hydrocarbons such as BTEX compounds (benzene, toluene, ethylbenzene and xylenes), and chlorinated solvents including tetrachloroethylene (PCE), trichloroethylene (TCE), and vinyl chloride (VC). BTEX are important components of gasoline. PCE and TCE are industrial solvents historically used in dry cleaning processes and as a metal degreaser, respectively.

Other organic pollutants present in groundwater and derived from industrial operations are the polycyclic aromatic hydrocarbons (PAHs). Due to its molecular weight, Naphthalene is the most soluble and mobile PAH found in groundwater, whereas benzo(a)pyrene is the most toxic one. PAHs are generally produced as byproducts by incomplete combustion of organic matter.

Organic pollutants can also be found in groundwater as insecticides and herbicides. As many other synthetic organic compounds, most pesticides have very complex molecular structures. This complexity determines the water solubility, adsorption capacity, and mobility of pesticides in the groundwater system. Thus, some types of pesticides are more mobile than others so they can more easily reach a drinking-water source.

Metals

Several trace metals occur naturally in certain rock formations and can enter in the environment from natural processes such as weathering. However, industrial activities such as mining, metallurgy, solid waste disposal, paint and enamel works, etc. can lead to elevated concentrations of toxic metals including lead, cadmium and chromium. These contaminants have the potential to make their way into groundwater.

The migration of metals (and metalloids) in groundwater will be affected by several factors, in particular by chemical reactions which determine the partitioning of contaminants among different phases and species. Thus, the mobility of metals primarily depends on the pH and redox state of groundwater.

Pharmaceuticals

Trace amounts of pharmaceuticals from treated wastewater infiltrating into the aquifer are among emerging ground-water contaminants being studied throughout the United States. Popular phar-maceuticals such as antibiotics, anti-inflammatories, antidepressants, decongestants, tranquiliz-ers, etc. are normally found in treated wastewater. This wastewater is discharged from the treat-ment facility, and often makes its way into the aquifer or source of surface water used for drinking water.

Trace amounts of pharmaceuticals in both groundwater and surface water are far below what is considered dangerous or of concern in most areas, but it could be an increasing problem as popu-lation grows and more reclaimed wastewater is utilized for municipal water supplies.

Others

Other organic pollutants include a range of organohalides and other chemical compounds, petro-leum hydrocarbons, various chemical compounds found in personal hygiene and cosmetic prod-ucts, drug pollution involving pharmaceutical drugs and their metabolites. Inorganic pollutants might include other nutrients such as ammonia and phosphate, and radionuclides such as urani-um (U) or radon (Rn) naturally present in some geological formations. Saltwater intrusion is also an example of natural contamination, but is very often intensified by human activities.

Groundwater pollution is a worldwide issue. A study of the groundwater quality of the principal aqui-fers of the United States conducted between 1991 and 2004, showed that 23% of domestic wells had contaminants at levels greater than human-health benchmarks. Another study suggested that the major groundwater pollution problems in Africa, considering the order of importance are: (1) nitrate pollution, (2) pathogenic agents, (3) organic pollution, (4) salinization, and (5) acid mine drainage.

Causes

Causes of groundwater pollution include:

- Naturally-occurring (geogenic)
- On-site sanitation systems
- Sewage (treated and untreated)
- Fertilizers and pesticides
- Commercial and industrial leaks
- Hydraulic fracturing
- Landfill leachate
- Other

Naturally-occurring (Geogenic)

"Geogenic" refers to naturally occurring as a result from geological processes.

The natural arsenic pollution occurs because aquifer sediments contain organic matter that gen-erates anaerobic conditions in the aquifer. These conditions result in the microbial dissolution of iron oxides in the sediment and, thus, the release of the arsenic, normally strongly bound to iron oxides, into the water. As a consequence, arsenic-rich groundwater is often iron-rich, although secondary processes often obscure the association of dissolved arsenic and dissolved iron. Arsenic is found in groundwater most commonly as the reduced species arsenite and the oxidized species arsenate, being the acute toxicity of arsenite somewhat greater than that of arsenate. Investigations by WHO indicated that 20% of 25,000 boreholes tested in Bangladesh had arsenic concentrations exceeding 50 µg/l.

The occurrence of fluoride is close related to the abundance and solubility of fluoride-containing minerals such as fluorite (CaF_2). Considerably high concentrations of fluoride in groundwater are typically caused by a lack of calcium in the aquifer. Health problems associated with dental fluorosis may occur when fluoride concentrations in groundwater exceed 1.5 mg/l, which is the WHO guideline value since 1984.

The Swiss Federal Institute of Aquatic Science and Technology (EAWAG) has recently developed the interactive Groundwater Assessment Platform (GAP), where the geogenic risk of contamination in a given area can be estimated using geological, topographical and other environmental data without having to test samples from every single groundwater resource. This tool also allows the user to produce probability risk mapping for both arsenic and fluoride.

High concentrations of parameters like salinity, iron, manganese, uranium, radon and chromium, in groundwater, may also be of geogenic origin. This contaminants can be important locally but they are not as widespread as arsenic and fluoride.

On-site Sanitation Systems

A traditional housing compound near Herat, Afghanistan, where a shallow water supply well (foreground) is in close proximity to the pit latrine (behind the white greenhouse) leading to contamination of the groundwater.

Groundwater pollution with pathogens and nitrate can also occur from the liquids infiltrating into the ground from on-site sanitation systems such as pit latrines and septic tanks, depending on the population density and the hydrogeological conditions.

Factors controlling the fate and transport of pathogens are quite complex and the interaction among them is not well understood. If the local hydrogeological conditions (which can vary within a space of a few square kilometres) are ignored, simple on-site sanitation infrastructures such as pit latrines can cause significant public health risks via contaminated groundwater.

Liquids leach from the pit and pass the unsaturated soil zone (which is not completely filled with water). Subsequently, these liquids from the pit enter the groundwater where they may lead to groundwater pollution. This is a problem if a nearby water well is used to supply groundwater for drinking water purposes. During the passage in the soil, pathogens can die off or be adsorbed significantly, mostly depending on the travel time between the pit and the well. Most, but not all pathogens die within 50 days of travel through the subsurface.

The degree of pathogen removal strongly varies with soil type, aquifer type, distance and other environmental factors. For example, the unsaturated zone becomes "washed" during extended periods of heavy rain, providing hydraulic pathway for the quick pass of pathogens. It is difficult to estimate the safe distance between a pit latrine or a septic tank and a water source. In any case, such recommendations about the safe distance are mostly ignored by those building pit latrines. In addition, household plots are of a limited size and therefore pit latrines are often built much closer to groundwater wells than what can be regarded as safe. This results in groundwater pollution and household members falling sick when using this groundwater as a source of drinking water.

Sewage (Treated and Untreated)

Groundwater pollution can be caused by untreated waste discharge leading to diseases like skin lesions, bloody diarrhea and dermatitis. This is more common in locations having limited wastewater treatment infrastructure, or where there are systematic failures of the on-site sewage disposal system. Along with pathogens and nutrients, untreated sewage can also have an important load of heavy metals that may seep into the groundwater system.

The treated effluent from sewage treatment plants may also reach the aquifer if the effluent is infiltrated or discharged to local surface water bodies. Therefore, those substances that are not removed in conventional sewage treatment plants may reach the groundwater as well. For example, detected concentrations of pharmaceutical residues in groundwater were in the order of 50 ng/L in several locations in Germany. This is because in conventional sewage treatment plants, micro-pollutants such as hormones, pharmaceutical residues and other micro-pollutants contained in urine and feces are only partially removed and the remainder is discharged into surface water, from where it may also reach the groundwater.

Groundwater pollution can also occur from leaking sewers which has been observed for example in Germany. This can also lead to potential cross-contamination of drinking-water supplies.

Spreading wastewater or sewage sludge in agriculture may also be included as sources of faecal contamination in groundwater.

Fertilizers and Pesticides

Nitrate can also enter the groundwater via excessive use of fertilizers, including manure spreading. This is because only a fraction of the nitrogen-based fertilizers is converted to produce and other plant matter. The remainder accumulates in the soil or lost as run-off. High application rates of nitrogen-containing fertilizers combined with the high water-solubility of nitrate leads to increased runoff into surface water as well as leaching into groundwater, thereby causing groundwater pollution. The excessive use of nitrogen-containing fertilizers (be they synthetic or natural) is particularly damaging, as much of the nitrogen that is not taken up by plants is transformed into nitrate which is easily leached.

Poor management practices in manure spreading can introduce both
pathogens and nutrients (nitrate) in the groundwater system.

The nutrients, especially nitrates, in fertilizers can cause problems for natural habitats and for human health if they are washed off soil into watercourses or leached through soil into groundwater. The heavy use of nitrogenous fertilizers in cropping systems is the largest contributor to anthropogenic nitrogen in groundwater worldwide.

Feedlots/animal corrals can also lead to the potential leaching of nitrogen and metals to groundwater. Over application of animal manure may also result in groundwater pollution with pharmaceutical residues derived from veterinary drugs.

The US Environmental Protection Agency (EPA)and the European Commission are seriously dealing with the nitrate problem related to agricultural development, as a major water supply problem that requires appropriate management and governance.

Runoff of pesticides may leach into groundwater causing human health problems from contaminated water wells. Pesticide concentrations found in groundwater are typically low, and often the regulatory human health-based limits exceeded are also very low. The organophosphorus insecticide monocrotophos (MCP) appears to be one of a few hazardous, persistent, soluble and mobile (it does not bind with minerals in soils) pesticides able to reach a drinking-water source. In general, more pesticide compounds are being detected as groundwater quality monitoring programs have become more extensive; however, much less monitoring has been conducted in developing countries due to the high analysis costs.

Commercial and Industrial Leaks

A wide variety of both inorganic and organic pollutants have been found in aquifers underlying commercial and industrial activities.

Ore mining and metal processing facilities are the primary responsible of the presence of metals in groundwater of anthropogenic origin, including arsenic. The low pH associated with acid mine drainage (AMD) contributes to the solubility of potential toxic metals that can eventually enter the groundwater system.

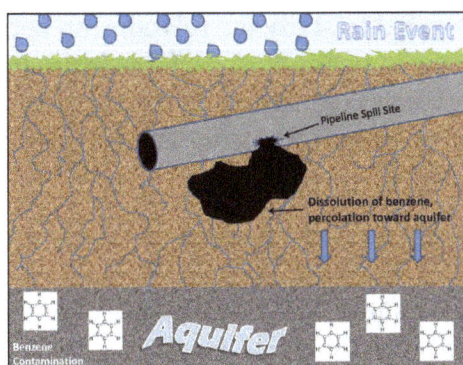

Oil spills associated with underground pipelines and tanks can release benzene and other soluble petroleum hydrocarbons that rapidly percolate down into the aquifer.

There is an increasing concern over the groundwater pollution by gasoline leaked from petroleum underground storage tanks (USTs) of gas stations. BTEX compounds are the most common additives of the gasoline. BTEX compounds, including benzene, have densities lower than water (1 g/ml). Similar to the oil spills on the sea, the non-miscible phase, referred to as Light Non-Aqueous Phase Liquid (LNAPL), will "float" upon the water table in the aquifer.

Chlorinated solvents are used in nearly any industrial practice where degreasing removers are required. PCE is a highly utilized solvent in the dry cleaning industry because of its cleaning effectiveness and relatively low cost. It has also been used for metal-degreasing operations. Because it is highly volatile, it is more frequently found in groundwater than in surface water. TCE has historically been used as a metal cleaning. The military facility Anniston Army Dept (ANAD) in the United States was placed on the EPA Superfund National Priorities List (NPL) because of groundwater contamination with as much as 27 million pounds of TCE. Both PCE and TCE may degrade to vinyl chloride (VC), the most toxic chlorinated hydrocarbon.

Many types of solvents may have also been disposed illegally, leaking over time to the groundwater system.

Chlorinated solvents such as PCE and TCE have densities higher than water and the non-miscible phase is referred to as Dense Non-Aqueous Phase Liquids (DNAPL). Once they reach the aquifer, they will "sink" and eventually accumulate on the top of low-permeability layers. Historically, wood-treating facilities have also release insecticides such as pentachlorophenol (PCP) and creosote into the environment, impacting the groundwater resources. PCP is a highly soluble and toxic obsolete pesticide recently listed in the Stockholm Convention on Persistent Organic Pollutants. PAHs and other semi-VOCs are the common contaminants associated with creosote.

Although non-miscible, both LNAPLs and DNAPLs still have the potential to slowly dissolve into the aqueous (miscible) phase to create a plume and thus become a long-term source of contamination. DNAPLs (chlorinated solvents, heavy PAHs, creosote, PCBs) tend to be difficult to manage as they can reside very deep in the groundwater system.

Hydraulic Fracturing

The recent growth of hydraulic fracturing ("Fracking") wells in the United States has raised concerns regarding its potential risks of contaminating groundwater resources. EPA, along with many other researchers, has been delegated to study the relationship between hydraulic fracturing and drinking water resources. While it is possible to perform hydraulic fracturing without having a relevant impact on groundwater resources if stringent controls and quality management measures are in place, there are a number of cases where groundwater pollution due to improper handling or technical failures was observed.

While the EPA has not found significant evidence of a widespread, systematic impact on drinking water by hydraulic fracturing, this may be due to insufficient systematic pre-and post-hydraulic fracturing data on drinking water quality, and the presence of other agents of contamination that preclude the link between tight oil and shale gas extraction and its impact.

Despite the EPA's lack of profound widespread evidence, other researchers have made significant observations of rising groundwater contamination in close proximity to major shale oil/gas drilling sites located in Marcellus (British Columbia, Canada). Within one kilometer of these specific sites, a subset of shallow drinking water consistently showed higher concentration levels of methane, ethane, and propane concentrations than normal. An evaluation of higher Helium and other noble gas concentration along with the rise of hydrocarbon levels supports the distinction between hydraulic fracturing fugitive gas and naturally occurring "background" hydrocarbon content. This contamination is speculated to be the result of leaky, failing, or improperly installed gas well casings.

Furthermore, it is theorized that contamination could also result from the capillary migration of deep residual hyper-saline water and hydraulic fracturing fluid, slowly flowing through faults and fractures until finally making contact with groundwater resources; however, many researchers argue that the permeability of rocks overlying shale formations are too low to allow this to ever happen sufficiently. To ultimately prove this theory, there would have to be traces of toxic trihalomethanes (THM) since they are often associated with the presence of stray gas contamination, and typically co-occur with high halogen concentrations in hyper-saline waters. Besides, highly saline waters are a common natural feature in deep groundwater systems.

While conclusions regarding groundwater pollution as the result to hydraulic fracturing fluid flow is restricted in both space and time, researchers have hypothesized that the potential for systematic stray gas contamination depends mainly on the integrity of the shale oil/gas well structure, along with its relative geological location to local fracture systems that could potentially provide flow paths for fugitive gas migration.

Though widespread, systematic contamination by hydraulic fracturing has been heavily disputed, one major source of contamination that has the most consensus among researchers of being the most problematic is site-specific accidental spillage of hydraulic fracturing fluid and produced water. So far, a significant majority of groundwater contamination events are derived from surface-level anthropogenic routes rather than the subsurface flow from underlying shale formations. While the damage can be obvious, and much more effort is being done to prevent these accidents from occurring so frequently, the lack of data from fracking oil spills continue to leave researchers in the dark. In many of these events, the data acquired from the leakage or spillage is often very vague, and thus would lead researchers to lacking conclusions.

Researchers from the Federal Institute for Geosciences and Natural Resources (BGR) conducted a modelling study for a deep shale-gas formation in the North German Basin. They concluded that the probability is small that the rise of fracking fluids through the geological underground to the surface will impact shallow groundwater.

Landfill Leachate

Leachate from sanitary landfills can lead to groundwater pollution.

Love Canal was one of the most widely known examples of groundwater pollution. In 1978, residents of the Love Canal neighborhood in upstate New York noticed high rates of cancer and an alarming number of birth defects. This was eventually traced to organic solvents and dioxins from an industrial landfill that the neighborhood had been built over and around, which had then infiltrated into the water supply and evaporated in basements to further contaminate the air. Eight hundred families were reimbursed for their homes and moved, after extensive legal battles and media coverage.

Other

Further causes of groundwater pollution are chemical spills from commercial or industrial operations, chemical spills occurring during transport (e.g. spillage of diesel fuels), illegal waste dumping, infiltration from urban runoff or mining operations, road salts, de-icing chemicals from airports and even atmospheric contaminants since groundwater is part of the hydrologic cycle.

The burial of corpses and their subsequent degradation may also pose a risk of pollution to groundwater.

Mechanisms

The passage of water through the subsurface can provide a reliable natural barrier to contamination but it only works under favorable conditions.

The stratigraphy of the area plays an important role in the transport of pollutants. An area can have layers of sandy soil, fractured bedrock, clay, or hardpan. Areas of karst topography on limestone bedrock are sometimes vulnerable to surface pollution from groundwater. Earthquake faults can also be entry routes for downward contaminant entry. Water table conditions are of great importance for drinking water supplies, agricultural irrigation, waste disposal (including nuclear waste), wildlife habitat, and other ecological issues.

Many chemicals undergo reactive decay or chemical change, especially over long periods of time in groundwater reservoirs. A noteworthy class of such chemicals is the chlorinated hydrocarbons such as trichloroethylene (used in industrial metal degreasing and electronics manufacturing) and tetrachloroethylene used in the dry cleaning industry. Both of these chemicals, which are carcinogens themselves, undergo partial decomposition reactions, leading to new hazardous chemicals (including dichloroethylene and vinyl chloride).

Interactions with surface water

Although interrelated, surface water and groundwater have often been studied and managed as separate resources. Surface water seeps through the soil and becomes groundwater. Conversely, groundwater can also feed surface water sources. Sources of surface water pollution are generally grouped into two categories based on their origin.

Interactions between groundwater and surface water are complex. Consequently, groundwater pollution, sometimes referred to as groundwater contamination, is not as easily classified as surface water pollution. By its very nature, groundwater aquifers are susceptible to contamination from sources that may not directly affect surface water bodies, and the distinction of point vs. non-point source may be irrelevant.

A spill or ongoing release of chemical or radionuclide contaminants into soil (located away from a surface water body) may not create point or non-point source pollution but can contaminate the aquifer below, creating a toxic plume. The movement of the plume, may be analyzed through a hydrological transport model or groundwater model.

Prevention

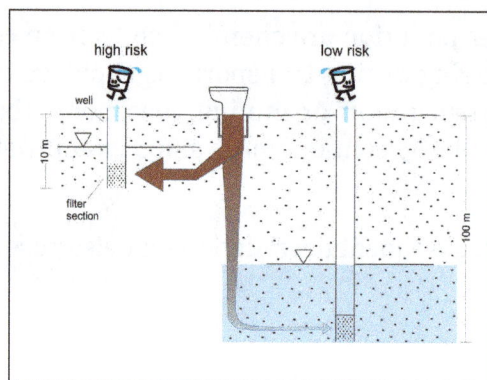

Schematic showing that there is a lower risk of groundwater pollution with greater depth of the water well.

Precautionary Principle

The precautionary principle, evolved from Principle 15 of the Rio Declaration on Environment and Development, is important in protecting groundwater resources from pollution. The precautionary principle provides that *"where there are threats of irreversible damage, lack of full scientific certainty shall not be used as reason for postponing cost-effective measures to prevent environmental degradation"*.

One of the six basic principles of the European Union (EU) water policy is the application of the precautionary principle.

Groundwater Quality Monitoring

Groundwater quality monitoring programs have been implemented regularly in many countries around the world. They are important components to understand the hydrogeological system, and for the development of conceptual models and aquifer vulnerability maps.

Groundwater quality must be regularly monitored across the aquifer to determine trends. Effective groundwater monitoring should be driven by a specific objective, for example, a specific contaminant of concern. Contaminant levels can be compared to the World Health Organization (WHO) guidelines for drinking-water quality. It is not rare that limits of contaminants are reduced as more medical experience is gained.

Sufficient investment should be given to continue monitoring over the long term. When a problem is found, action should be taken to correct it. Waterborne outbreaks in the United States decreased with the introduction of more stringent monitoring (and treatment) requirements in the early 90s.

The community can also help monitor the groundwater quality:

Land Zoning for Groundwater Protection

The development of land-use zoning maps has been implemented by several water authorities at different scales around the world. There are two types of zoning maps: aquifer vulnerability maps and source protection maps.

Aquifer Vulnerability Map

It refers to the intrinsic (or natural) vulnerability of a groundwater system to pollution. Intrinsically, some aquifers are more vulnerable to pollution than other aquifers. Shallow unconfined aquifers are more at risk of pollution because there are fewer layers to filter out contaminants.

The unsaturated zone can play an important role in retarding (and in some cases eliminating) pathogens and so must be considered when assessing aquifer vulnerability. The biological activity is greatest in the top soil layers where the attenuation of pathogens is generally most effective.

Preparation of the vulnerability maps typically involves overlaying several thematic maps of physical factors that have been selected to describe the aquifer vulnerability. The index-based parametric mapping method GOD developed by Foster and Hirata uses three generally available or readily estimated parameters, the degree of Groundwater hydraulic confinement, geological nature of the Overlying strata and Depth to groundwater. A further approach developed by EPA, a rating system named "DRASTIC," employs seven hydrogeological factors to develop an index of vulnerability: Depth to water table, net Recharge, Aquifer media, Soil media, Topography (slope), Impact on the vadose zone, and hydraulic Conductivity.

There is a particular debate among hydrogeologists as to whether aquifer vulnerability should be established in a general (intrinsic) way for all contaminants, or specifically for each pollutant.

Source Protection Map

It refers to the capture areas around an individual groundwater source, such as a water well or a spring, to especially protect them from pollution. Thus, potential sources of degradable pollutants, such as pathogens, can be located at distances which travel times along the flowpaths are long enough for the pollutant to be eliminated through filtration or adsorption.

Analytical methods using equations to define groundwater flow and contaminant transport are the most widely used. The WHPA is a semi-analytical groundwater flow simulation program developed by the US EPA for delineating capture zones in a wellhead protection area.

The simplest form of zoning employs fixed-distance methods where activities are excluded within a uniformly applied specified distance around abstraction points.

Locating on-site Sanitation Systems

As the health effects of most toxic chemicals arise after prolonged exposure, risk to health from chemicals is generally lower than that from pathogens. Thus, the quality of the source protection measures is an important component in controlling whether pathogens may be present in the final drinking-water.

On-site sanitation systems can be designed in such a way that groundwater pollution from these sanitation systems is prevented from occurring. Detailed guidelines have been developed to estimate safe distances to protect groundwater sources from pollution from on-site sanitation. The following criteria have been proposed for safe siting (i.e. deciding on the location) of on-site sanitation systems:

- Horizontal distance between the drinking water source and the sanitation system:

    ○ Guideline values for horizontal separation distances between on-site sanitation systems and water sources vary widely (e.g. 15 to 100 m horizontal distance between pit latrine and groundwater wells).

- Vertical distance between drinking water well and sanitation system.

- Aquifer type.

- Groundwater flow direction.

- Impermeable layers.

- Slope and surface drainage.

- Volume of leaking wastewater.

- Superposition, i.e. the need to consider a larger planning area.

As a very general guideline it is recommended that the bottom of the pit should be at least 2 m above groundwater level, and a minimum horizontal distance of 30 m between a pit and a water source is normally recommended to limit exposure to microbial contamination. However, no general statement should be made regarding the minimum lateral separation distances required to prevent contamination of a well from a pit latrine. For example, even 50 m lateral separation distance might not be sufficient in a strongly karstified system with a downgradient supply well or spring, while 10 m lateral separation distance is completely sufficient if there is a well developed clay cover layer and the annular space of the groundwater well is well sealed.

Legislation

Institutional and legal issues are critical in determining the success or failure of groundwater protection policies and strategies.

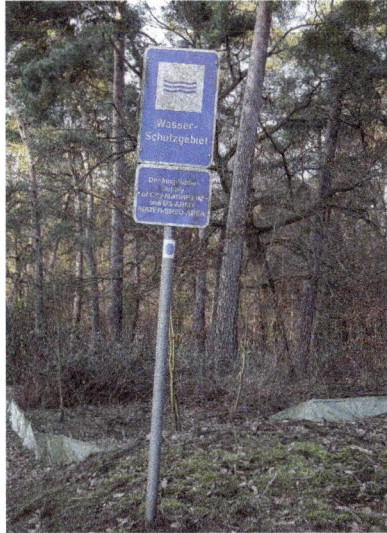

Sign near Mannheim, Germany indicating a zone as a dedicated "groundwater protection zone".

United States

In November 2006, EPA published its *Ground Water Rule*, due to concerns that public water systems supplied by ground water would be vulnerable to contamination from harmful microorganisms, including fecal matter. The objective of the regulation, promulgated under the authority of the Safe Drinking Water Act, is to keep microbial pathogens out of public water sources.

Management

Options for remediation of contaminated groundwater can be grouped into the following categories:

- Containing the pollutants to prevent them from migrating further.
- Removing the pollutants from the aquifer.
- Remediating the aquifer by either immobilizing or detoxifying the contaminants while they are still in the aquifer (in-situ).
- Treating the groundwater at its point of use.
- Abandoning the use of this aquifer's groundwater and finding an alternative source of water.

Point-of-use Treatment

Portable water purification devices or "point-of-use" (POU) water treatment systems and field water disinfection techniques can be used to remove some forms of groundwater pollution prior to drinking, namely any fecal pollution. Many commercial portable water purification systems or chemical additives are available which can remove pathogens, chlorine, bad taste, odors, and heavy metals like lead and mercury.

Techniques include boiling, filtration, activated charcoal absorption, chemical disinfection, ultraviolet purification, ozone water disinfection, solar water disinfection, solar distillation, homemade water filters.

Arsenic removal filters (ARF) are dedicated technologies typically installed to remove arsenic. Many of these technologies require a capital investment and long-term maintenance. Filters in Bangladesh are usually abandoned by the users due to their high cost and complicated maintenance, which is also quite expensive.

Groundwater Remediation

Groundwater pollution is much more difficult to abate than surface pollution because groundwater can move great distances through unseen aquifers. Non-porous aquifers such as clays partially purify water of bacteria by simple filtration (adsorption and absorption), dilution, and, in some cases, chemical reactions and biological activity; however, in some cases, the pollutants merely transform to soil contaminants. Groundwater that moves through open fractures and caverns is not filtered and can be transported as easily as surface water. In fact, this can be aggravated by the human tendency to use natural sinkholes as dumps in areas of karst topography.

Pollutants and contaminants can be removed from ground water by applying various techniques thereby making it safe for use. Ground water treatment (or remediation) techniques span biological, chemical, and physical treatment technologies. Most ground water treatment techniques utilize a combination of technologies. Some of the biological treatment techniques include bioaugmentation, bioventing, biosparging, bioslurping, and phytoremediation. Some chemical treatment techniques include ozone and oxygen gas injection, chemical precipitation, membrane separation, ion exchange, carbon absorption, aqueous chemical oxidation, and surfactant-enhanced recovery. Some chemical techniques may be implemented using nanomaterials. Physical treatment techniques include, but are not limited to, pump and treat, air sparging, and dual phase extraction.

Abandonment

If treatment or remediation of the polluted groundwater is deemed to be too difficult or expensive, then abandoning the use of this aquifer's groundwater and finding an alternative source of water is the only other option.

Oil Spill Water Pollution

An oil spill is the release of a liquid petroleum hydrocarbon into the environment, especially the marine ecosystem, due to human activity, and is a form of pollution. The term is usually given to marine oil spills, where oil is released into the ocean or coastal waters, but spills may also occur on land. Oil spills may be due to releases of crude oil from tankers, offshore platforms, drilling rigs and wells, as well as spills of refined petroleum products (such as gasoline, diesel) and their by-products, heavier fuels used by large ships such as bunker fuel, or the spill of any oily refuse or waste oil.

Oil spills penetrate into the structure of the plumage of birds and the fur of mammals, reducing its insulating ability, and making them more vulnerable to temperature fluctuations and much less buoyant in the water. Cleanup and recovery from an oil spill is difficult and depends upon many factors, including the type of oil spilled, the temperature of the water (affecting evaporation and biodegradation), and the types of shorelines and beaches involved. Spills may take weeks, months or even years to clean up.

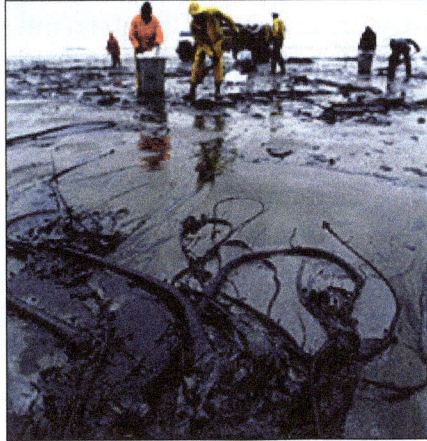

Kelp after an oil spill.

Oil spills can have disastrous consequences for society; economically, environmentally, and socially. As a result, oil spill accidents have initiated intense media attention and political uproar, bringing many together in a political struggle concerning government response to oil spills and what actions can best prevent them from happening.

Oil slick from the Montara oil spill in the Timor Sea, September 2009.

Human Impact

An oil spill represents an immediate fire hazard. The Kuwaiti oil fires produced air pollution that caused respiratory distress. The *Deepwater Horizon* explosion killed eleven oil rig workers. The fire resulting from the Lac-Mégantic derailment killed 47 and destroyed half of the town's centre.

Spilled oil can also contaminate drinking water supplies. For example, in 2013 two different oil spills contaminated water supplies for 300,000 in Miri, Malaysia; 80,000 people in Coca, Ecuador. In 2000, springs were contaminated by an oil spill in Clark County, Kentucky.

Contamination can have an economic impact on tourism and marine resource extraction industries. For example, the *Deepwater Horizon* oil spill impacted beach tourism and fishing along the Gulf Coast, and the responsible parties were required to compensate economic victims.

Environmental Effects

In general, spilled oil can affect animals and plants in two ways: direct from the oil and from the response or cleanup process. There is no clear relationship between the amount of oil in the aquatic environment and the likely impact on biodiversity. A smaller spill at the wrong time/wrong season and in a sensitive environment may prove much more harmful than a larger spill at another time of the year in another or even the same environment. Oil penetrates into the structure of the plumage of birds and the fur of mammals, reducing their insulating ability, and making them more vulnerable to temperature fluctuations and much less buoyant in the water.

A surf scoter covered in oil as a result of the 2007 San Francisco Bay oil spill.

A bird covered in oil from the Black Sea oil spill.

Animals who rely on scent to find their babies or mothers cannot due to the strong scent of the oil. This causes a baby to be rejected and abandoned, leaving the babies to starve and eventually die. Oil can impair a bird's ability to fly, preventing it from foraging or escaping from predators. As they preen, birds may ingest the oil coating their feathers, irritating the digestive tract, altering liver function, and causing kidney damage. Together with their diminished foraging capacity, this can rapidly result in dehydration and metabolic imbalance. Some birds exposed to petroleum also experience changes in their hormonal balance, including changes in their luteinizing protein. The majority of birds affected by oil spills die from complications without human intervention. Some studies have suggested that less than one percent of oil-soaked birds survive, even after cleaning, although the survival rate can also exceed ninety percent, as in the case of the Treasure oil spill.

Heavily furred marine mammals exposed to oil spills are affected in similar ways. Oil coats the fur of sea otters and seals, reducing its insulating effect, and leading to fluctuations in body temperature and hypothermia. Oil can also blind an animal, leaving it defenseless. The ingestion of oil causes dehydration and impairs the digestive process. Animals can be poisoned, and may die from oil entering the lungs or liver.

There are three kinds of oil-consuming bacteria. Sulfate-reducing bacteria (SRB) and acid-producing bacteria are anaerobic, while general aerobic bacteria (GAB) are aerobic. These bacteria occur naturally and will act to remove oil from an ecosystem, and their biomass will tend to replace other populations in the food chain. The chemicals from the oil which dissolve in water, and hence are available to bacteria, are those in the water associated fraction of the oil.

In addition, oil spills can also harm air quality. The chemicals in crude oil are mostly hydrocarbons that contains toxic chemicals such as benzenes, toluene, poly-aromatic hydrocarbon and oxygenated polycyclic aromatic hydrocarbons. These chemicals can introduce adverse health effects when being inhaled into human body. In addition, these chemicals can be oxidized by oxidants in the atmosphere to form fine particulate matter after they evaporate into the atmosphere. These particulates can penetrate lungs and carry toxic chemicals into the human body. Burning surface oil can also be a source for pollution such as soot particles. During the cleanup and recovery process, it will also generate air pollutants such as nitric oxides and ozone from ships. Lastly, bubble bursting can also be a generation pathway for particulate matter during an oil spill. During the Deepwater Horizon oil spill, significant air quality issues were found on the Gulf Coast, which is the downwind of DWH oil spill. Air quality monitoring data showed that criteria pollutants had exceeded the health-based standard in the coastal regions.

Sources and Rate of Occurrence

A VLCC tanker can carry 2 million barrels (320,000 m³) of crude oil. This is about eight times the amount spilled in the widely known *Exxon Valdez* oil spill. In this spill, the ship ran aground and dumped 260,000 barrels (41,000 m³) of oil into the ocean in March 1989. Despite efforts of scientists, managers, and volunteers over 400,000 seabirds, about 1,000 sea otters, and immense numbers of fish were killed. Considering the volume of oil carried by sea, however, tanker owners' organisations often argue that the industry's safety record is excellent, with only a tiny fraction of a percentage of oil cargoes carried ever being spilled. The International Association of Independent Tanker Owners has observed that "accidental oil spills this decade have been at record low levels— one third of the previous decade and one tenth of the 1970s—at a time when oil transported has more than doubled since the mid 1980s."

Oil tankers are just one of the many sources of oil spills. According to the United States Coast Guard, 35.7% of the volume of oil spilled in the United States from 1991 to 2004 came from tank vessels (ships/barges), 27.6% from facilities and other non-vessels, 19.9% from non-tank vessels, and 9.3% from pipelines; 7.4% from mystery spills. On the other hand, only 5% of the actual spills came from oil tankers, while 51.8% came from other kinds of vessels.

The International Tanker Owners Pollution Federation has tracked 9,351 accidental spills that have occurred since 1974. According to this study, most spills result from routine operations such as loading cargo, discharging cargo, and taking on fuel oil. 91% of the operational oil spills are small, resulting in less than 7 metric tons per spill. On the other hand, spills resulting from accidents like collisions, groundings, hull failures, and explosions are much larger, with 84% of these involving losses of over 700 metric tons.

Cleanup and Recovery

A U.S. Air Force Reserve plane sprays Corexit dispersant over the Deepwater Horizon oil spill in the Gulf of Mexico.

Clean-up efforts after the Exxon Valdez oil spill.

A US Navy oil spill response team drills with a "Harbour Buster high-speed oil containment system".

Cleanup and recovery from an oil spill is difficult and depends upon many factors, including the type of oil spilled, the temperature of the water (affecting evaporation and biodegradation), and the types of shorelines and beaches involved. Physical cleanups of oil spills are also very expensive. However, microorganisms such as Fusobacteria species demonstrate an innovative potential for future oil spill cleanup because of their ability to colonize and degrade oil slicks on the sea surface.

Methods for cleaning up include:

- Bioremediation: Use of microorganisms or biological agents to break down or remove oil; such as the bacteria Alcanivorax or Methylocella Silvestris.

Oil slicks on Lake Maracaibo.

- Bioremediation Accelerator: Oleophilic, hydrophobic chemical, containing no bacteria, which chemically and physically bonds to both soluble and insoluble hydrocarbons. The bioremediation accelerator acts as a herding agent in water and on the surface, floating molecules to the surface of the water, including solubles such as phenols and BTEX, forming gel-like agglomerations. Undetectable levels of hydrocarbons can be obtained in produced water and manageable water columns. By overspraying sheen with bioremediation accelerator, sheen is eliminated within minutes. Whether applied on land or on water, the nutrient-rich emulsion creates a bloom of local, indigenous, pre-existing, hydrocarbon-consuming bacteria. Those specific bacteria break down the hydrocarbons into water and carbon dioxide, with EPA tests showing 98% of alkanes biodegraded in 28 days; and aromatics being biodegraded 200 times faster than in nature they also sometimes use the hydrofireboom to clean the oil up by taking it away from most of the oil and burning it.

- Controlled burning can effectively reduce the amount of oil in water, if done properly. But it can only be done in low wind, and can cause air pollution.

Volunteers cleaning up the aftermath of the Prestige oil spill.

- Dispersants can be used to dissipate oil slicks. A dispersant is either a non-surface active polymer or a surface-active substance added to a suspension, usually a colloid, to improve the separation of particles and to prevent settling or clumping. They may rapidly disperse large amounts of certain oil types from the sea surface by transferring it into the water column. They will cause the oil slick to break up and form water-soluble micelles that are rapidly diluted. The oil is then effectively spread throughout a larger volume of water than the surface from where the oil was dispersed. They can also delay the formation of persistent oil-in-water emulsions. However, laboratory experiments showed that dispersants increased toxic hydrocarbon levels in fish by a factor of up to 100 and may kill fish eggs. Dispersed oil droplets infiltrate into deeper water and can lethally contaminate coral. Research indicates that some dispersants are toxic to corals. A 2012 study found that Corexit dispersant had increased the toxicity of oil by up to 52 times.

- Watch and wait: In some cases, natural attenuation of oil may be most appropriate, due to the invasive nature of facilitated methods of remediation, particularly in ecologically sensitive areas such as wetlands.

- Dredging: For oils dispersed with detergents and other oils denser than water.

- Skimming: Requires calm waters at all times during the process.

- Solidifying: Solidifiers are composed of tiny, floating, dry ice pellets, and hydrophobic polymers that both adsorb and absorb. They clean up oil spills by changing the physical state of spilled oil from liquid to a solid, semi-solid or a rubber-like material that floats on water. Solidifiers are insoluble in water, therefore the removal of the solidified oil is easy and the oil will not leach out. Solidifiers have been proven to be relatively non-toxic to aquatic and wild life and have been proven to suppress harmful vapors commonly associated with hydrocarbons such as benzene, xylene and naphtha. The reaction time for solidification of oil is controlled by the surface area or size of the polymer or dry pellets as well as the viscosity and thickness of the oil layer. Some solidifier product manufactures claim the solidified oil can be thawed and used if frozen with dry ice or disposed of in landfills, recycled as an additive in asphalt or rubber products, or burned as a low ash fuel. A solidifier called C.I.Agent (manufactured by C.I.Agent Solutions of Louisville, Kentucky) is being used by BP in granular form, as well as in Marine and Sheen Booms at Dauphin Island and Fort Morgan, Alabama, to aid in the Deepwater Horizon oil spill cleanup.

- Vacuum and centrifuge: Oil can be sucked up along with the water, and then a centrifuge can be used to separate the oil from the water – allowing a tanker to be filled with near pure oil. Usually, the water is returned to the sea, making the process more efficient, but allowing small amounts of oil to go back as well. This issue has hampered the use of centrifuges due to a United States regulation limiting the amount of oil in water returned to the sea.

- Beach Raking: Coagulated oil that is left on the beach can be picked up by machinery.

Bags of oily waste from the *Exxon Valdez* oil spill.

Equipment used includes:

- Booms: Large floating barriers that round up oil and lift the oil off the water.

- Skimmers: Skim the oil.

- Sorbents: Large absorbents that absorb oil.

- Chemical and biological agents: Helps to break down the oil.

- Vacuums: Remove oil from beaches and water surface.

- Shovels and other road equipment: typically used to clean up oil on beaches.

Prevention

- Secondary containment – Methods to prevent releases of oil or hydrocarbons into environment.

- Oil Spill Prevention Control and Countermeasures (SPCC) program by the United States Environmental Protection Agency.

- Double-hulling – Build double hulls into vessels, which reduces the risk and severity of a spill in case of a collision or grounding. Existing single-hull vessels can also be rebuilt to have a double hull.

- Thick-hulled railroad transport tanks.

Spill response procedures should include elements such as:

- A listing of appropriate protective clothing, safety equipment, and cleanup materials required.

For spill cleanup (gloves, respirators, etc.) and an explanation of their proper use:

- Appropriate evacuation zones and procedures.

- Availability of fire suppression equipment.

- Disposal containers for spill cleanup materials.

- The first aid procedures that might be required.

Environmental Sensitivity Index (ESI) Mapping

Environmental Sensitivity Index (ESI) maps are used to identify sensitive shoreline resources prior to an oil spill event in order to set priorities for protection and plan cleanup strategies. By planning spill response ahead of time, the impact on the environment can be minimized or prevented. Environmental sensitivity index maps are basically made up of information within the following three categories: shoreline type, and biological and human-use resources.

Shoreline Type

Shoreline type is classified by rank depending on how easy the target site would be to clean up, how long the oil would persist, and how sensitive the shoreline is. The floating oil slicks put the shoreline at particular risk when they eventually come ashore, covering the substrate with oil. The differing substrates between shoreline types vary in their response to oiling, and influence the type of cleanup that will be required to effectively decontaminate the shoreline. In 1995, the US National Oceanic and Atmospheric Administration extended ESI maps to lakes, rivers, and estuary shoreline types. The exposure the shoreline has to wave energy and tides, substrate type, and slope of the shoreline are also taken into account—in addition to biological productivity and sensitivity. The productivity of the shoreline habitat is also taken into account when determining ESI ranking. Mangroves and marshes tend to have higher ESI rankings due to the potentially long-lasting and damaging effects of both the oil contamination and cleanup actions. Impermeable and exposed surfaces with high wave action are ranked lower due to the reflecting waves keeping oil from coming onshore, and the speed at which natural processes will remove the oil.

Biological Resources

Habitats of plants and animals that may be at risk from oil spills are referred to as "elements" and are divided by functional group. Further classification divides each element into species groups with similar life histories and behaviors relative to their vulnerability to oil spills. There are eight element groups: Birds, Reptiles, Amphibians, Fish, Invertebrates, Habitats and Plants, Wetlands, and Marine Mammals and Terrestrial Mammals. Element groups are further divided into subgroups, for example, the 'marine mammals' element group is divided into dolphins, manatees, pinnipeds (seals, sea lions & walruses), polar bears, sea otters and whales. Problems taken into consideration when ranking biological resources include the observance of a large number of individuals in a small area, whether special life stages occur ashore (nesting or molting), and whether there are species present that are threatened, endangered or rare.

Human-use Resources

Human use resources are divided into four major classifications; archaeological importance or cultural resource site, high-use recreational areas or shoreline access points, important protected management areas, or resource origins. Some examples include airports, diving sites, popular beach sites, marinas, natural reserves or marine sanctuaries.

Estimating the Volume of a Spill

By observing the thickness of the film of oil and its appearance on the surface of the water, it is possible to estimate the quantity of oil spilled. If the surface area of the spill is also known, the total volume of the oil can be calculated.

Appearance	Film thickness			Quantity spread	
	inches	mm	nm	gal/sq mi	L/ha
Barely visible	0.0000015	0.0000380	38	25	0.370
Silvery sheen	0.0000030	0.0000760	76	50	0.730
First trace of color	0.0000060	0.0001500	150	100	1.500
Bright bands of color	0.0000120	0.0003000	300	200	2.900
Colors begin to dull	0.0000400	0.0010000	1000	666	9.700
Colors are much darker	0.0000800	0.0020000	2000	1332	19.500

Oil spill model systems are used by industry and government to assist in planning and emergency decision making. Of critical importance for the skill of the oil spill model prediction is the adequate description of the wind and current fields. There is a worldwide oil spill modelling (WOSM) program. Tracking the scope of an oil spill may also involve verifying that hydrocarbons collected during an ongoing spill are derived from the active spill or some other source. This can involve sophisticated analytical chemistry focused on finger printing an oil source based on the complex mixture of substances present. Largely, these will be various hydrocarbons, among the most useful being polyaromatic hydrocarbons. In addition, both oxygen and nitrogen heterocyclic hydrocarbons, such as parent and alkyl homologues of carbazole, quinoline, and pyridine, are present in many crude oils. As a result, these compounds have great potential to supplement the existing suite of hydrocarbons targets to fine-tune source tracking of petroleum spills. Such analysis can also be used to follow weathering and degradation of crude spills.

Chemical Water Pollution

Ever since the industrial age, chemical water pollution is one of the most known and witnessed, of all the different types of water pollution. When foreign and unwanted chemicals are added to water bodies, it leads to chemical water pollution. Such incidents can occur due to the spilling of chemicals by accident, chemical pesticides from farms into streams, factory wastes etc.

The presence of toxic chemicals in the water harms not just the marine wildlife and eco-system, but also humans and other animals, who may end up consuming this water. Water contaminated due to chemical pollution is known to cause serious health issues.

Industrial and agricultural work involves the use of many different chemicals that can run-off into water and pollute it.

Metals and solvents from industrial work can pollute rivers and lakes. These are poisonous to many forms of aquatic life and may slow their development, make them infertile or even result in death.

Pesticides are used in farming to control weeds, insects and fungi. Run-offs of these pesticides can cause water pollution and poison aquatic life. Subsequently, birds, humans and other animals may be poisoned if they eat infected fish.

Marine Pollution

Marine pollution occurs when harmful effects result from the entry into the ocean of chemicals, particles, industrial, agricultural, and residential waste, noise, or the spread of invasive organisms. Eighty percent of marine pollution comes from land. Air pollution is also a contributing factor by carrying off pesticides or dirt into the ocean. Land and air pollution have proven to be harmful to marine life and its habitats.

The pollution often comes from nonpoint sources such as agricultural runoff, wind-blown debris, and dust. Nutrient pollution, a form of water pollution, refers to contamination by excessive inputs of nutrients. It is a primary cause of eutrophication of surface waters, in which excess nutrients, usually nitrates or phosphates, stimulate algae growth. Many potentially toxic chemicals adhere to tiny particles which are then taken up by plankton and benthic animals, most of which are either deposit feeders or filter feeders. In this way, the toxins are concentrated upward within ocean food chains. Many particles combine chemically in a manner highly depletive of oxygen, causing estuaries to become anoxic.

When pesticides are incorporated into the marine ecosystem, they quickly become absorbed into marine food webs. Once in the food webs, these pesticides can cause mutations, as well as diseases, which can be harmful to humans as well as the entire food web. Toxic metals can also be introduced into marine food webs. These can cause a change to tissue matter, biochemistry, behaviour, reproduction, and suppress growth in marine life. Also, many animal feeds have a high fish meal or fish hydrolysate content. In this way, marine toxins can be transferred to land animals, and appear later in meat and dairy products.

In order to protect the ocean from marine pollution, policies have been developed internationally. There are different ways for the ocean to get polluted, therefore there have been multiple laws, policies, and treaties put into place throughout history.

While marine pollution can be obvious, as with the marine debris shown above,
it is often the pollutants that cannot be seen that cause most harm.

Pathways of Pollution

Septic river.

There are many ways to categorize and examine the inputs of pollution into our marine ecosystems. Patin (n.d.) notes that generally there are three main types of inputs of pollution into the ocean: direct discharge of waste into the oceans, runoff into the waters due to rain, and pollutants released from the atmosphere.

One common path of entry by contaminants to the sea are rivers. The evaporation of water from oceans exceeds precipitation. The balance is restored by rain over the continents entering rivers and then being returned to the sea. The Hudson in New York State and the Raritan in New Jersey, which empty at the northern and southern ends of Staten Island, are a source of mercury contamination of zooplankton (copepods) in the open ocean. The highest concentration in the filter-feeding copepods is not at the mouths of these rivers but 70 miles (110 km) south, nearer Atlantic City, because water flows close to the coast. It takes a few days before toxins are taken up by the plankton.

Pollution is often classed as point source or nonpoint source pollution. Point source pollution occurs when there is a single, identifiable, localized source of the pollution. An example is directly discharging sewage and industrial waste into the ocean. Pollution such as this occurs particularly in developing nations. Nonpoint source pollution occurs when the pollution comes from ill-defined and diffuse sources. These can be difficult to regulate. Agricultural runoff and wind blown debris are prime examples.

Direct Discharge

Acid mine drainage in the Rio Tinto River.

Pollutants enter rivers and the sea directly from urban sewerage and industrial waste discharges, sometimes in the form of hazardous and toxic wastes.

Inland mining for copper, gold, etc., is another source of marine pollution. Most of the pollution is simply soil, which ends up in rivers flowing to the sea. However, some minerals discharged in the course of the mining can cause problems, such as copper, a common industrial pollutant, which can interfere with the life history and development of coral polyps. Mining has a poor environmental track record. For example, according to the United States Environmental Protection Agency, mining has contaminated portions of the headwaters of over 40% of watersheds in the western continental US. Much of this pollution finishes up in the sea.

Land Runoff

Surface runoff from farming, as well as urban runoff and runoff from the construction of roads, buildings, ports, channels, and harbours, can carry soil and particles laden with carbon, nitrogen, phosphorus, and minerals. This nutrient-rich water can cause fleshy algae and phytoplankton to thrive in coastal areas; known as algal blooms, which have the potential to create hypoxic conditions by using all available oxygen. In the coast of southwest Florida, harmful algal blooms have existed for over 100 years. These algal blooms have been a cause of species of fish, turtles, dolphins, and shrimp to die and cause harmful effects on humans who swim in the water.

Polluted runoff from roads and highways can be a significant source of water pollution in coastal areas. About 75% of the toxic chemicals that flow into Puget Sound are carried by stormwater that runs off paved roads and driveways, rooftops, yards and other developed land. In California, there are many rainstorms that runoff into the ocean. These rainstorms occur from October to March, and these runoff waters contain petroleum, heavy metals, pollutants from emissions, etc.

In China, there is a large coastal population that pollutes the ocean through land runoff. This includes sewage discharge and pollution from urbanization and land use. In 2001, more than 66,795 mi^2 of the Chinese coastal ocean waters were rated less than Class I of the Sea Water Quality Standard of China. Much of this pollution came from Ag, Cu, Cd, Pb, As, DDT, PCBs, etc., which occurred from contamination through land runoff.

Ship Pollution

A cargo ship pumps ballast water over the side.

Ships can pollute waterways and oceans in many ways. Oil spills can have devastating effects. While being toxic to marine life, polycyclic aromatic hydrocarbons (PAHs), found in crude oil, are very difficult to clean up, and last for years in the sediment and marine environment.

Oil spills are probably the most emotive of marine pollution events. However, while a tanker wreck may result in extensive newspaper headlines, much of the oil in the world's seas comes from other smaller sources, such as tankers discharging ballast water from oil tanks used on return ships, leaking pipelines or engine oil disposed of down sewers.

Discharge of cargo residues from bulk carriers can pollute ports, waterways, and oceans. In many instances vessels intentionally discharge illegal wastes despite foreign and domestic regulation prohibiting such actions. An absence of national standards provides an incentive for some cruise liners to dump waste in places where the penalties are inadequate. It has been estimated that container ships lose over 10,000 containers at sea each year (usually during storms). Ships also create noise pollution that disturbs natural wildlife, and water from ballast tanks can spread harmful algae and other invasive species.

Ballast water taken up at sea and released in port is a major source of unwanted exotic marine life. The invasive freshwater zebra mussels, native to the Black, Caspian, and Azov seas, were probably transported to the Great Lakes via ballast water from a transoceanic vessel. Meinesz believes that one of the worst cases of a single invasive species causing harm to an ecosystem can be attributed to a seemingly harmless jellyfish. *Mnemiopsis leidyi*, a species of comb jellyfish that spread so it now inhabits estuaries in many parts of the world. It was first introduced in 1982, and thought to have been transported to the Black Sea in a ship's ballast water. The population of the jellyfish grew exponentially and, by 1988, it was wreaking havoc upon the local fishing industry. "The anchovy catch fell from 204,000 tons in 1984 to 200 tons in 1993; sprat from 24,600 tons in 1984 to 12,000 tons in 1993; horse mackerel from 4,000 tons in 1984 to zero in 1993." Now that the jellyfish have exhausted the zooplankton, including fish larvae, their numbers have fallen dramatically, yet they continue to maintain a stranglehold on the ecosystem.

Invasive species can take over once occupied areas, facilitate the spread of new diseases, introduce new genetic material, alter underwater seascapes, and jeopardize the ability of native species to obtain food. Invasive species are responsible for about $138 billion annually in lost revenue and management costs in the US alone.

Atmospheric Pollution

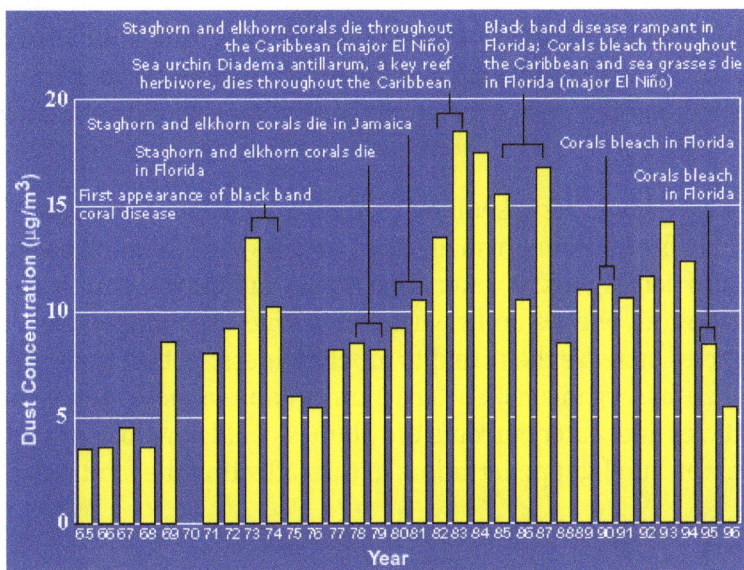

A graph linking atmospheric dust to various coral deaths across the Caribbean Sea and Florida.

Another pathway of pollution occurs through the atmosphere. Wind-blown dust and debris, including plastic bags, are blown seaward from landfills and other areas. Dust from the Sahara moving around the southern periphery of the subtropical ridge moves into the Caribbean and Florida during the warm season as the ridge builds and moves northward through the subtropical Atlantic. Dust can also be attributed to a global transport from the Gobi and Taklamakan deserts across Korea, Japan, and the Northern Pacific to the Hawaiian Islands.

Since 1970, dust outbreaks have worsened due to periods of drought in Africa. There is a large variability in dust transport to the Caribbean and Florida from year to year; however, the flux is greater during positive phases of the North Atlantic Oscillation. The USGS links dust events to a decline in the health of coral reefs across the Caribbean and Florida, primarily since the 1970s.

Climate change is raising ocean temperatures and raising levels of carbon dioxide in the atmosphere. These rising levels of carbon dioxide are acidifying the oceans. This, in turn, is altering aquatic ecosystems and modifying fish distributions, with impacts on the sustainability of fisheries and the livelihoods of the communities that depend on them. Healthy ocean ecosystems are also important for the mitigation of climate change.

Deep Sea Mining

Deep sea mining is a relatively new mineral retrieval process that takes place on the ocean floor. Ocean mining sites are usually around large areas of polymetallic nodules or active and extinct hydrothermal vents at about 1,400 – 3,700 meters below the ocean's surface. The vents create sulfide deposits, which contain precious metals such as silver, gold, copper, manganese, cobalt, and zinc. The deposits are mined using either hydraulic pumps or bucket systems that take ore to the surface to be processed. As with all mining operations, deep sea mining raises questions about environmental damages to the surrounding areas.

Because deep sea mining is a relatively new field, the complete consequences of full-scale mining operations are unknown. However, experts are certain that removal of parts of the sea floor will result in disturbances to the benthic layer, increased toxicity of the water column, and sediment plumes from tailings. Removing parts of the sea floor disturbs the habitat of benthic organisms, possibly, depending on the type of mining and location, causing permanent disturbances. Aside from direct impact of mining the area, leakage, spills, and corrosion would alter the mining area's chemical makeup.

Among the impacts of deep sea mining, sediment plumes could have the greatest impact. Plumes are caused when the tailings from mining (usually fine particles) are dumped back into the ocean, creating a cloud of particles floating in the water. Two types of plumes occur: near-bottom plumes and surface plumes. Near-bottom plumes occur when the tailings are pumped back down to the mining site. The floating particles increase the turbidity, or cloudiness, of the water, clogging filter-feeding apparatuses used by benthic organisms. Surface plumes cause a more serious problem. Depending on the size of the particles and water currents the plumes could spread over vast areas. The plumes could impact zooplankton and light penetration, in turn affecting the food web of the area.

Types of Pollution

Acidification

An island with a fringing reef in the Maldives. Coral reefs are dying around the world.

The oceans are normally a natural carbon sink, absorbing carbon dioxide from the atmosphere. Because the levels of atmospheric carbon dioxide are increasing, the oceans are becoming more acidic. The potential consequences of ocean acidification are not fully understood, but there are concerns that structures made of calcium carbonate may become vulnerable to dissolution, affecting corals and the ability of shellfish to form shells.

Oceans and coastal ecosystems play an important role in the global carbon cycle and have removed about 25% of the carbon dioxide emitted by human activities between 2000 and 2007 and about half the anthropogenic CO_2 released since the start of the industrial revolution. Rising ocean temperatures and ocean acidification means that the capacity of the ocean carbon sink will gradually get weaker, giving rise to global concerns expressed in the Monaco and Manado Declarations.

A report from NOAA scientists published in the journal Science in May 2008 found that large amounts of relatively acidified water are upwelling to within four miles of the Pacific continental shelf area of North America. This area is a critical zone where most local marine life lives or is born. While the paper dealt only with areas from Vancouver to northern California, other continental shelf areas may be experiencing similar effects.

A related issue is the methane clathrate reservoirs found under sediments on the ocean floors. These trap large amounts of the greenhouse gas methane, which ocean warming has the potential to release. In 2004 the global inventory of ocean methane clathrates was estimated to occupy between one and five million cubic kilometres. If all these clathrates were to be spread uniformly across the ocean floor, this would translate to a thickness between three and fourteen metres. This estimate corresponds to 500–2500 gigatonnes carbon (Gt C), and can be compared with the 5000 Gt C estimated for all other fossil fuel reserves.

Eutrophication

A polluted lagoon.

Eutrophication is an increase in chemical nutrients, typically compounds containing nitrogen or phosphorus, in an ecosystem. It can result in an increase in the ecosystem's primary productivity (excessive plant growth and decay), and further effects including lack of oxygen and severe reductions in water quality, fish, and other animal populations.

The biggest culprit are rivers that empty into the ocean, and with it the many chemicals used as fertilizers in agriculture as well as waste from livestock and humans. An excess of oxygen-depleting chemicals in the water can lead to hypoxia and the creation of a dead zone.

Estuaries tend to be naturally eutrophic because land-derived nutrients are concentrated where runoff enters the marine environment in a confined channel. The World Resources Institute has identified 375 hypoxic coastal zones around the world, concentrated in coastal areas in Western

Europe, the Eastern and Southern coasts of the US, and East Asia, particularly in Japan. In the ocean, there are frequent red tide algae blooms that kill fish and marine mammals and cause respiratory problems in humans and some domestic animals when the blooms reach close to shore.

In addition to land runoff, atmospheric anthropogenic fixed nitrogen can enter the open ocean. A study in 2008 found that this could account for around one third of the ocean's external (non-recycled) nitrogen supply and up to three per cent of the annual new marine biological production. It has been suggested that accumulating reactive nitrogen in the environment may have consequences as serious as putting carbon dioxide in the atmosphere.

One proposed solution to eutrophication in estuaries is to restore shellfish populations, such as oysters. Oyster reefs remove nitrogen from the water column and filter out suspended solids, subsequently reducing the likelihood or extent of harmful algal blooms or anoxic conditions. Filter feeding activity is considered beneficial to water quality by controlling phytoplankton density and sequestering nutrients, which can be removed from the system through shellfish harvest, buried in the sediments, or lost through denitrification. Foundational work toward the idea of improving marine water quality through shellfish cultivation to was conducted by Odd Lindahl et al., using mussels in Sweden.

Plastic Debris

A mute swan builds a nest using plastic garbage.

Marine debris is mainly discarded human rubbish which floats on, or is suspended in the ocean. Eighty percent of marine debris is plastic – a component that has been rapidly accumulating since the end of World War II. The mass of plastic in the oceans may be as high as 100,000,000 tonnes (98,000,000 long tons; 110,000,000 short tons).

Discarded plastic bags, six pack rings, cigarette butts and other forms of plastic waste which finish up in the ocean present dangers to wildlife and fisheries. Aquatic life can be threatened through entanglement, suffocation, and ingestion. Fishing nets, usually made of plastic, can be left or lost in the ocean by fishermen. Known as ghost nets, these entangle fish, dolphins, sea turtles, sharks, dugongs, crocodiles, seabirds, crabs, and other creatures, restricting movement, causing starvation, laceration, infection, and, in those that need to return to the surface to breathe, suffocation.

Many animals that live on or in the sea consume flotsam by mistake, as it often looks similar to their natural prey. Plastic debris, when bulky or tangled, is difficult to pass, and may become permanently lodged in the digestive tracts of these animals. Especially when evolutionary adaptions make it impossible for the likes of turtles to reject plastic bags, which resemble jellyfish when immersed in water, as they have a system in their throat to stop slippery foods from otherwise escaping. Thereby blocking the passage of food and causing death through starvation or infection.

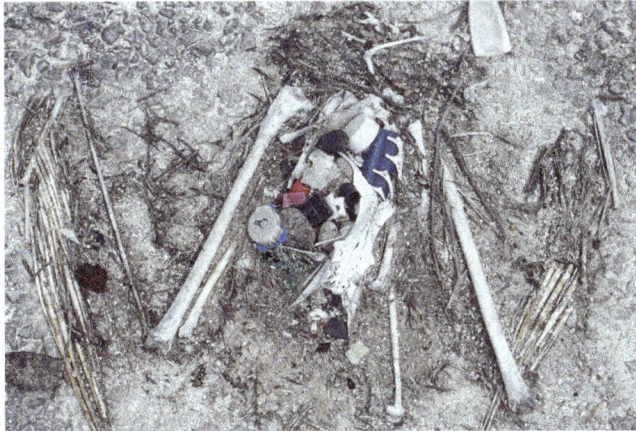
The remains of an albatross containing ingested flotsam.

Plastics accumulate because they don't biodegrade in the way many other substances do. They will photodegrade on exposure to the sun, but they do so properly only under dry conditions, and water inhibits this process. In marine environments, photodegraded plastic disintegrates into ever-smaller pieces while remaining polymers, even down to the molecular level. When floating plastic particles photodegrade down to zooplankton sizes, jellyfish attempt to consume them, and in this way the plastic enters the ocean food chain.

Many of these long-lasting pieces end up in the stomachs of marine birds and animals, including sea turtles, and black-footed albatross. In a 2008 Pacific Gyre voyage, Algalita Marine Research Foundation researchers began finding that fish are ingesting plastic fragments and debris. Of the 672 fish caught during that voyage, 35% had ingested plastic pieces.

Plastic debris tends to accumulate at the centre of ocean gyres. The North Pacific Gyre, for example, has collected the so-called "Great Pacific Garbage Patch", which is now estimated to be one to twenty times the size of Texas (approximately from 700,000 to 15,000,000 square kilometers). There could be as much plastic as fish in the sea. It has a very high level of plastic particulate suspended in the upper water column. In samples taken in 1999, the mass of plastic exceeded that of zooplankton (the dominant animal life in the area) by a factor of six.

Midway Atoll, in common with all the Hawaiian Islands, receives substantial amounts of debris from the garbage patch. Ninety percent plastic, this debris accumulates on the beaches of Midway where it becomes a hazard to the bird population of the island. Midway Atoll is home to two-thirds (1.5 million) of the global population of Laysan albatross. Nearly all of these albatross have plastic in their digestive system and one-third of their chicks die.

Toxic additives used in the manufacture of plastic materials can leach out into their surroundings when exposed to water. Waterborne hydrophobic pollutants collect and magnify on the surface of plastic debris, thus making plastic far more deadly in the ocean than it would be on land. Hydrophobic contaminants are also known to bioaccumulate in fatty tissues, biomagnifying up the food chain and putting pressure on apex predators. Some plastic additives are known to disrupt the endocrine system when consumed, others can suppress the immune system or decrease reproductive rates.

Floating debris can also absorb persistent organic pollutants from seawater, including PCBs, DDT, and PAHs. Aside from toxic effects, when ingested some of these affect animal brain cells similarly to estradiol, causing hormone disruption in the affected wildlife. Saido, a chemist with the College of Pharmacy, conducted a study in Nihon University, Chiba, Japan, that discovered, when plastics eventually decompose, they produce potentially toxic bisphenol A (BPA) and PS oligomer into the water. These toxins are believed to bring harm to the marine life living in the area.

A growing concern regarding plastic pollution in the marine ecosystem is the use of microplastics. Microplastics are little beads of plastic less than 5 millimeters wide, and they are commonly found in hand soaps, face cleansers, and other exfoliators. When these products are used, the microplastics go through the water filtration system and into the ocean, but because of their small size they are likely to escape capture by the preliminary treatment screens on wastewater plants. These beads are harmful to the organisms in the ocean, especially filter feeders, because they can easily ingest the plastic and become sick. The microplastics are such a concern because it is difficult to clean them up due to their size, so humans can try to avoid using these harmful plastics by purchasing products that use environmentally safe exfoliates.

Toxins

Apart from plastics, there are particular problems with other toxins that do not disintegrate rapidly in the marine environment. Examples of persistent toxins are PCBs, DDT, TBT, pesticides, furans, dioxins, phenols, and radioactive waste. Heavy metals are metallic chemical elements that have a relatively high density and are toxic or poisonous at low concentrations. Examples are mercury, lead, nickel, arsenic, and cadmium. Such toxins can accumulate in the tissues of many species of aquatic life in a process called bioaccumulation. They are also known to accumulate in benthic environments, such as estuaries and bay muds: a geological record of human activities of the last century.

Specific examples:

- Chinese and Russian industrial pollution such as phenols and heavy metals in the Amur River have devastated fish stocks and damaged its estuary soil.

- Wabamun Lake in Alberta, Canada, once the best whitefish lake in the area, now has unacceptable levels of heavy metals in its sediment and fish.

- Acute and chronic pollution events have been shown to impact southern California kelp forests, though the intensity of the impact seems to depend on both the nature of the contaminants and duration of exposure.

- Due to their high position in the food chain and the subsequent accumulation of heavy metals from their diet, mercury levels can be high in larger species such as bluefin and albacore. As a result, in March 2004 the United States FDA issued guidelines recommending that pregnant women, nursing mothers and children limit their intake of tuna and other types of predatory fish.

- Some shellfish and crabs can survive polluted environments, accumulating heavy metals or toxins in their tissues. For example, mitten crabs have a remarkable ability to survive in highly modified aquatic habitats, including polluted waters. The farming and harvesting of such species needs careful management if they are to be used as a food.

- Surface runoff of pesticides can alter the gender of fish species genetically, transforming male into female fish.

- Heavy metals enter the environment through oil spills – such as the Prestige oil spill on the Galician coast and Gulf of Mexico which unleashed an estimated 3.19 million barrels of oil – or from other natural or anthropogenic sources.

- In 2005, the 'Ndrangheta, an Italian mafia syndicate, was accused of sinking at least 30 ships loaded with toxic waste, much of it radioactive. This has led to widespread investigations into radioactive waste disposal rackets.

- Since the end of World War II, various nations, including the Soviet Union, the United Kingdom, the United States, and Germany, have disposed of chemical weapons in the Baltic Sea, raising concerns of environmental contamination.

- The damaging of the Fukushima Dai-ichi Nuclear Power Plant in 2011 caused radioactive toxins to leak into the air and ocean. There are still many isotopes in the ocean, which directly affects the benthic food web and also affects the whole food chain. The concentration of 137Cs in the bottom sediment that was contaminated by water with high concentrations in April–May 2011 remains quite high and is showing signs of very slow decrease with time.

Underwater Noise

Marine life can be susceptible to noise or the sound pollution from sources such as passing ships, oil exploration seismic surveys, and naval low-frequency active sonar. Sound travels more rapidly and over larger distances in the sea than in the atmosphere. Marine animals, such as cetaceans, often have weak eyesight, and live in a world largely defined by acoustic information. This applies also to many deeper sea fish, who live in a world of darkness. Between 1950 and 1975, ambient noise at one location in the Pacific Ocean increased by about ten decibels (that is a tenfold increase in intensity).

Noise also makes species communicate louder, which is called the Lombard vocal response. Whale songs are longer when submarine-detectors are on. If creatures don't "speak" loud enough, their voice can be masked by anthropogenic sounds. These unheard voices might be warnings, finding of prey, or preparations of net-bubbling. When one species begins speaking louder, it will mask other species voices, causing the whole ecosystem to eventually speak louder.

According to the oceanographer Sylvia Earle, "Undersea noise pollution is like the death of a thousand cuts. Each sound in itself may not be a matter of critical concern, but taken all together, the noise from shipping, seismic surveys, and military activity is creating a totally different environment than existed even 50 years ago. That high level of noise is bound to have a hard, sweeping impact on life in the sea."

Noise from ships and human activity can damage Cnidarians and Ctenophora, which are very important organisms in the marine ecosystem. They promote high diversity and they are used as models for ecology and biology because of their simple structures. When there is underwater noise, the vibrations in the water damage the cilia hairs in the Coelenterates. In a study, the organisms were exposed to sound waves for different numbers of times and the results showed that damaged hair cells were extruded or missing or presented bent, flaccid or missed kinocilia and stereocilia.

Adaptation and Mitigation

Aerosol can pollute beaches.

Much anthropogenic pollution ends up in the ocean. The 2011 edition of the United Nations Environment Programme Year Book identifies as the main emerging environmental issues the loss to the oceans of massive amounts of phosphorus, "a valuable fertilizer needed to feed a growing global population", and the impact billions of pieces of plastic waste are having globally on the health of marine environments.

Bjorn Jennssen notes in his article, "Anthropogenic pollution may reduce biodiversity and productivity of marine ecosystems, resulting in reduction and depletion of human marine food resources". There are two ways the overall level of this pollution can be mitigated: either the human population is reduced, or a way is found to reduce the ecological footprint left behind by the average human. If the second way is not adopted, then the first way may be imposed as the world ecosystems falter.

The second way is for humans, individually, to pollute less. That requires social and political will, together with a shift in awareness so more people respect the environment and are less disposed to abuse it. At an operational level, regulations, and international government participation is needed. It is often very difficult to regulate marine pollution because pollution spreads over international barriers, thus making regulations hard to create as well as enforce.

Without appropriate awareness of marine pollution, the necessary global will to effectively address the issues may prove inadequate. Balanced information on the sources and harmful effects of marine pollution need to become part of general public awareness, and ongoing research is required to fully establish, and keep current, the scope of the issues. As expressed in Daoji and Dag's research, one of the reasons why environmental concern is lacking among the Chinese is because the public awareness is low and therefore should be targeted.

The amount of awareness on marine pollution is vital to the support of keeping the prevention of trash from entering waterways and ending up in our oceans. The EPA reports that in 2014

Americans generated about 258 million tons of waste, and only a third was recycled or composted. In 2015, there was over 8 million tons of plastic that made it into the ocean. Through more sustainable packing this could lead to; eliminating toxic constituents, using fewer materials, making more readily available recyclable plastic. However, awareness can only take these initiatives so far. The most abundant plastic is PET (Polyethylene terephthalate) and is the most resistant to biodegradables. Researchers have been making great strides in combating this problem. In one way has been by adding a special polymer called a tetrablock copolymer. The tetrablock copolymer acts as a laminate between the PE and iPP which enables for an easier breakdown but still be tough. Through more awareness, individuals will become more cognizant of their carbon footprints. Also, from research and technology, more strides can be made to aid in the plastic pollution problem.

In 2018 a survey of Global Oceanic Environmental Survey (GOES) Foundation find that the ecosystem in seas and oceans can collapse in the next 25 years what would fail the terrestrial ecosystem and end the life on earth as we know them. The main causes: Plastic pollution, Ocean acidification, Ocean pollution. For prevent that scenario from happen, we need a total single use plastic ban, ban on wood burning, planting as many trees as possible, " pollution-free recycling of electronics and by 2030 all industries to be zero toxic discharge." "special protection and perservation of peat bogs, wetlands, marshlands and mangrove swamps to ensure carbon dioxide is absorbed from the atmosphere.

Plastic Resin Pellet Pollution

Plastic resin pellet pollution is a type of marine debris originating from plastic particles utilized in manufacturing large-scale plastics. These pre-production plastic pellets are created separately from the user plastics they are melted down to form, and pellet loss is incurred during both the manufacturing and transport stages. Commonly referred to as nurdles, these plastics are released into the open environment, creating pollution in the oceans and on beaches.

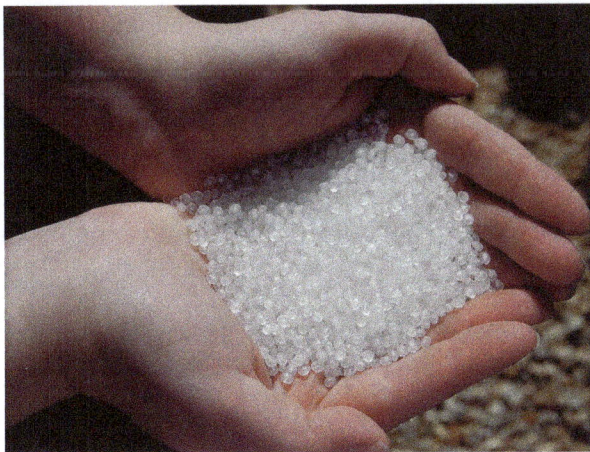

A handful of nurdles, spilled from a train in Pineville, Louisiana, in the United States.

Plastic resin pellets are classified as primary source microplastics, meaning that they were intentionally produced at the size ranging from 1–5 mm in diameter. Approximately 60 billion pounds (27 million tonnes) of nurdles are manufactured annually in the United States. One pound of pelletized HDPE contains approximately 25,000 nurdles (approximately 20 mg per nurdle). They are typically under 5 mm (0.20 in) in diameter.

Environmental Impact

Nurdles are a major contributor to marine debris. During a three-month study of Orange County beaches researchers found them to be the most common beach contaminant. Bathing beaches in East Lothian, Scotland have been shown to have covered with E. coli and Vibrio biofilms, according to a 2019 study.

Nurdles on a beach in southwest France, 2011.

Waterborne nurdles may either be a raw material of plastic production, or from larger chunks of plastics. A major concentration of plastic may be the Great Pacific garbage patch, a growing collection of marine debris known for its high concentrations of plastic litter.

Nurdles that escape from the plastic production process into waterways or oceans have become a significant source of ocean and beach plastic pollution. Plastic pellet pollution that has been monitored in studies is mainly found in the sediments and beach areas and is usually polyethylene or polypropylene, the two main plastic polymers found in microplastic pollution. Marine life is severely threatened by these small pieces of plastic; the creatures that make up the base of the marine food chain, such as krill, are prematurely dying by choking on nurdles.

Nurdles have frequently been found in the digestive tracts of various marine creatures, causing physiological damage by leaching plasticizers such as phthalates. Nurdles can carry two types of micropollutants in the marine environment: native plastic additives and hydrophobic pollutants absorbed from seawater. For example, concentrations of PCBs and DDE on nurdles collected from Japanese coastal waters were found to be up to 1 million times higher than the levels detected in surrounding seawater.

Plastic microbeads used in cosmetic exfoliating products are also found in water.

Incidents

2018

- USA - Pennsylvania - Semi-truck crash leading to release of bright blue colored nurdles into Pocono Creek, and the waterways of the Lehigh Valley.

2017

- South Africa - Durban - Nurdle spill of about 2 billion nurdles (49 tons), from a shipping container in Durban Harbor, requiring extended cleanup efforts. "The South African Association for Marine Biological Research (SAAMBR) has sent out an urgent appeal for beach users along the entire South African coast to try and assist in collecting as many of these nurdles as possible." These nurdles have also been spotted washing up on the shore in Western Australia.

- UK - "The Great Nurdle Hunt" occurring from June 2-5th, 2017 - across the United Kingdom drew attention to the issue of plastic pellet pollution. A program started by Fidra, a Scottish environmental charity, sourced information on nurdles from citizens across the region using shared photos to better understand the makeup of pollution across beaches in the UK. The nurdle hunts occurring earlier in 2017 determined that 73% of UK beaches had nurdle pollution.

2012

- USA - San Francisco Bay Coastal Cleanup from multiple nurdle spills.

- In Hong Kong, after being blown by Typhoon Vicente on 24 July 2012, some containers belonging to Chinese oil giant Sinopec which were carrying over 150 tonnes of plastic pellets were blown into the sea, washing up on southern Hong Kong coasts, such as Shek O, Cheung Chau, Ma Wan and Lamma Island. The spill disrupted marine life and is being credited with killing stocks of fish on fish farms.

Current Progress and Solutions

The plastic industry has responded to the increased interest and concern for plastic pellet loss and pollution sources. Operation Clean Sweep was created by SPI: The Plastics Industry Trade Association in 2001 and joined by the American Chemistry Council with the goal of zero pellet loss for plastic manufacturers. This voluntary stewardship program provides its members with a manual which guides them through ways in which they can reduce pellet loss within their own facilities and provides the necessary training.

In 2008, California passed a "nurdle law," which "specifically names pre-production plastic pellets (nurdles) as a pollutant".

Actions for Creating Awareness

On April 11, 2013 in order to create awareness, artist Maria Cristina Finucci founded The Garbage Patch State at UNESCO –Paris in front of Director General Irina Bokova. It is the first of a series of events under the patronage of UNESCO and of Italian Ministry of the Environment.

Eutrophication

Eutrophication or hypertrophication, is when a body of water becomes overly enriched with minerals and nutrients which induce excessive growth of algae. This process may result in oxygen depletion of the water body. One example is an "algal bloom" or great increase of phytoplankton in a water body as a response to increased levels of nutrients. Eutrophication is often induced by the discharge of nitrate or phosphate-containing detergents, fertilizers, or sewage into an aquatic system.

The eutrophication of the Potomac River is evident from the bright green water, caused by a dense bloom of cyanobacteria.

The eutrophication of the Potomac River is evident from the bright green water, caused by a dense bloom of cyanobacteria.

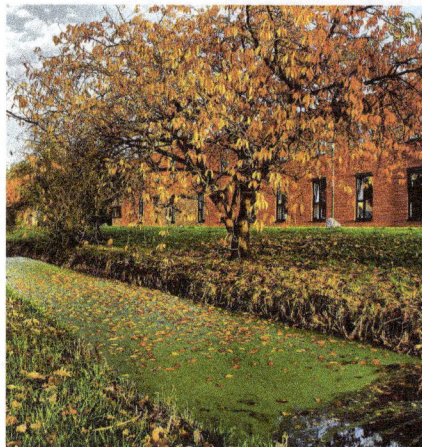

Eutrophication in a canal.

Mechanism of Eutrophication

Eutrophication most commonly arises from the oversupply of nutrients, most commonly as nitrogen or phosphorus, which leads to overgrowth of plants and algae in aquatic ecosystems. After such organisms die, bacterial degradation of their biomass results in oxygen consumption, thereby creating the state of hypoxia.

According to Ullmann's Encyclopedia, "the primary limiting factor for eutrophication is phosphate." The availability of phosphorus generally promotes excessive plant growth and decay, favouring simple algae and plankton over other more complicated plants, and causes a severe reduction in water quality. Phosphorus is a necessary nutrient for plants to live, and is the limiting factor for plant growth in many freshwater ecosystems. Phosphate adheres tightly to soil, so it is mainly transported by erosion. Once translocated to lakes, the extraction of phosphate into water is slow, hence the difficulty of reversing the effects of eutrophication. However, numerous literature report that nitrogen is the primary limiting nutrient for the accumulation of algal biomass.

The sources of these excess phosphates are phosphates in detergent, industrial/domestic run-offs, and fertilizers. With the phasing out of phosphate-containing detergents in the 1970s, industrial/domestic run-off and agriculture have emerged as the dominant contributors to eutrophication.

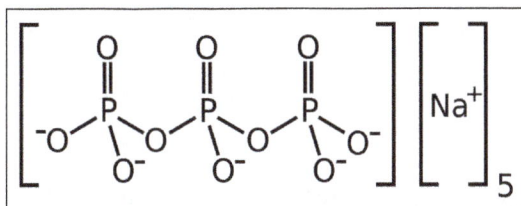

Sodium triphosphate, once a component of many detergents, was a major contributor to eutrophication.

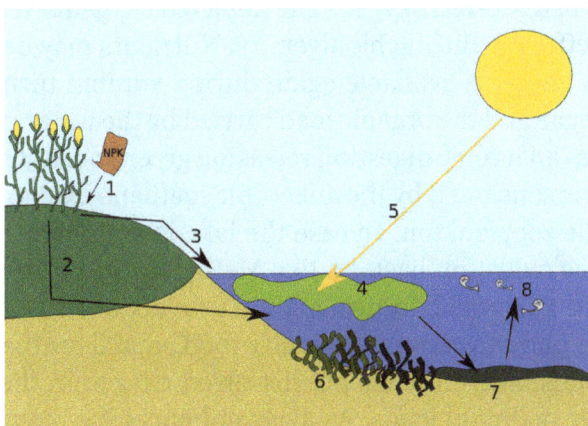

In the figure, 1. Excess nutrients are applied to the soil. 2. Some nutrients leach into the soil where they can remain for years. Eventually, they get drained into the water body. 3. Some nutrients run off over the ground into the body of water. 4. The excess nutrients cause an algal bloom. 5. The algal bloom blocks the light of the sun from reaching the bottom of the water body. 6. The plants beneath the algal bloom die because they cannot get sunlight to photosynthesize. 7. Eventually, the algal bloom dies and sinks to the bottom of the lake. Bacteria begins to decompose the remains, using up oxygen for respiration. 8. The decomposition causes the water to become depleted of oxygen. Larger life forms, such as fish, suffocate to death. This body of water can no longer support life.

Cultural Eutrophication

Cultural eutrophication is the process that speeds up natural eutrophication because of human activity. Due to clearing of land and building of towns and cities, land runoff is accelerated and more nutrients such as phosphates and nitrate are supplied to lakes and rivers, and then to coastal estuaries and bays. Extra nutrients are also supplied by treatment plants, golf courses, fertilizers, farms (including fish farms), as well as untreated sewage in many countries.

Lakes and Rivers

The eutrophication of the Mono Lake which is a cyanobacteria rich Soda lake.

When algae die, they decompose and the nutrients contained in that organic matter are converted into inorganic form by microorganisms. This decomposition process consumes oxygen, which reduces the concentration of dissolved oxygen. The depleted oxygen levels in turn may lead to fish kills and a range of other effects reducing biodiversity. Nutrients may become concentrated in an anoxic zone and may only be made available again during autumn turn-over or in conditions of turbulent flow. The dead algae and the organic load carried by the water inflows in to the lake settle at its bottom and undergoes anaerobic digestion releasing greenhouse gases like methane and CO_2. Some part of methane gas is consumed by the anaerobic methane oxidation bacteria which in turn works as food source to the zooplankton. In case the lake is not deficit of dissolved oxygen at all depths the aerobic methane oxidation bacteria like Methylococcus capsulatus can consume most of the methane by releasing CO_2 which in turn aid the production of algae. Thus a self-sustaining biological process can take place to generate primary food source for the phytoplankton and zooplankton depending on availability of adequate dissolved oxygen in the water bodies which are subjected to higher organic pollution loads. As algae enhances the dissolved oxygen by releasing oxygen from photosynthesis during the sunshine and consume oxygen by emitting CO_2 from its respiration during the absence of sunlight, adequate dissolved oxygen availability in water bodies is very crucial for fisheries production and elimination of green house gas emissions especially during the absence of sunlight in eutrophic water bodies. The CO_2 released by the algae during the absence of sunlight is stored in the water by reducing the water alkalinity and pH for its use during the sunshine.

Enhanced growth of aquatic vegetation or phytoplankton and algal blooms disrupts normal functioning of the ecosystem, causing a variety of problems such as a lack of oxygen needed for fish and shellfish to survive. The water becomes cloudy, typically coloured a shade of green, yellow, brown, or red. Eutrophication also decreases the value of rivers, lakes and aesthetic enjoyment. Health problems can occur where eutrophic conditions interfere with drinking water treatment.

Human activities can accelerate the rate at which nutrients enter ecosystems. Runoff from agriculture and development, pollution from septic systems and sewers, sewage sludge spreading, and other human-related activities increase the flow of both inorganic nutrients and organic substances into ecosystems. Elevated levels of atmospheric compounds of nitrogen can increase nitrogen availability. Phosphorus is often regarded as the main culprit in cases of eutrophication in lakes subjected to "point source" pollution from sewage pipes. The concentration of algae and the trophic state of lakes correspond well to phosphorus levels in water. Studies conducted in the Experimental Lakes Area in Ontario have shown a relationship between the addition of phosphorus and the rate of eutrophication. Humankind has increased the rate of phosphorus cycling on Earth by four times, mainly due to agricultural fertilizer production and application. Between 1950 and 1995, an estimated 600,000,000 tonnes of phosphorus was applied to Earth's surface, primarily on croplands.

Nutrient Pollution

Nutrient pollution, a form of water pollution, refers to contamination by excessive inputs of nutrients. It is a primary cause of eutrophication of surface waters, in which excess nutrients, usually nitrogen or phosphorus, stimulate algal growth. Sources of nutrient pollution include surface runoff from farm fields and pastures, discharges from septic tanks and feedlots, and emissions from combustion. Excess nutrients have been summarized as potentially leading to:

- Population effects: Excess growth of algae (blooms);

- Community effects: Species composition shifts (dominant taxa);

- Ecological effects: Food web changes, light limitation;

- Biogeochemical effects: Excess organic carbon (eutrophication); dissolved oxygen deficits (environmental hypoxia); toxin production;

- Human health effects: excess nitrate in drinking water (blue baby syndrome); disinfection by-products in drinking water;

- Biodiversity effects: Excessive algae blooms (biodiversity loss).

In a 2011 United States Environmental Protection Agency (EPA) report, the agency's Science Advisory Board succinctly stated: "Excess reactive nitrogen compounds in the environment are associated with many large-scale environmental concerns, including eutrophication of surface waters, toxic algae blooms, hypoxia, acid rain, nitrogen saturation in forests, and global warming."

Nutrient pollution caused by Surface runoff of soil and fertilizer during a rain storm.

Sources

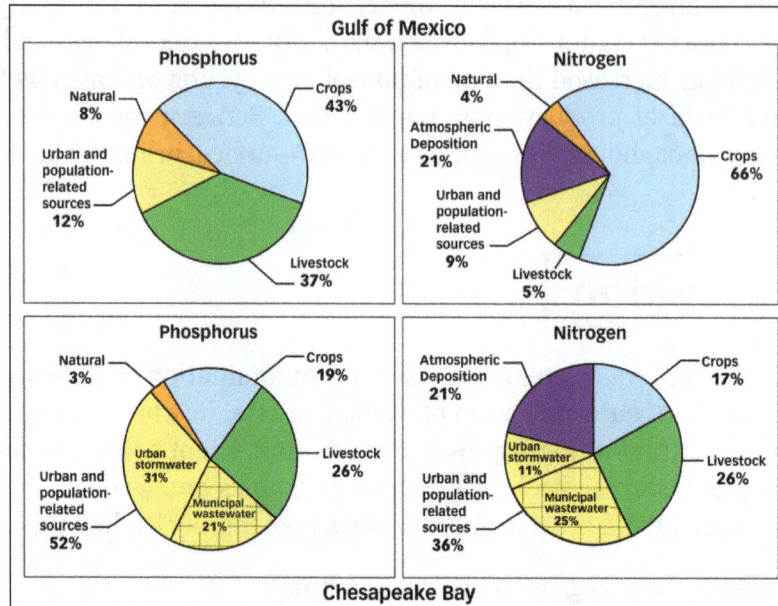

Agriculture is the major source of nutrient pollution in the Gulf of Mexico. In the Chesapeake Bay, agriculture is a major source, along with urban areas and atmospheric deposition.

Nitrogen

Use of synthetic fertilizers, burning of fossil fuels, and agricultural animal production, especially concentrated animal feeding operations (CAFO), have added large quantities of reactive nitrogen to the biosphere.

Phosphorus

Phosphorus pollution is caused by excessive use of fertilizers and manure, particularly when compounded by soil erosion. Phosphorus is also discharged by municipal sewage treatment plants and some industries.

Land Uses

The principal sources of nutrient pollution in an individual watershed depend on the prevailing land uses. The sources may be point sources, nonpoint sources, or both:

- Agriculture: Animal production or crops.

- Urban/suburban: Stormwater runoff from roads and parking lots; excessive fertilzer use on lawns; municipal sewage treatment plants; motor vehicle emissions.

- Industrial: Air pollution emissions (e.g. electric power plants), wastewater discharges from various industries.

Nutrient pollution from some air pollution sources may occur independently of the local land uses, due to long-range transport of air pollutants from distant sources.

Mitigation of Nutrient Pollutant Discharges

United States

Agricultural nonpoint source (NPS) pollution is the largest source of water quality impairments throughout the U.S., based on surveys by state environmental agencies. NPS pollution is not subject to discharge permits under the federal Clean Water Act (CWA). EPA and states have used grants, partnerships and demonstration projects to create incentives for farmers to adjust their practices and reduce surface runoff.

Discharge Permits

Many point source dischargers in the U.S., while not necessarily the largest sources of nutrients in their respective watersheds, are required to comply with nutrient effluent limitations in their permits, which are issued through the National Pollutant Discharge Elimination System (NPDES), pursuant to the CWA. Some large municipal sewage treatment plants, such as the Blue Plains Advanced Wastewater Treatment Plant in Washington, D.C. have installed biological nutrient removal (BNR) systems to comply with regulatory requirements. Other municipalities have made adjustments to the operational practices of their existing secondary treatment systems to control nutrients.

Discharges from large livestock facilities (CAFO) are also regulated by NPDES permits. Surface runoff from farm fields, the principal source of nutrients in many watersheds, is classified as NPS pollution and is not regulated by NPDES permits.

TMDL Program

A Total Maximum Daily Load (TMDL) is a regulatory plan that prescribes the maximum amount of a pollutant (including nutrients) that a body of water can receive while still meeting CWA water quality standards. Specifically, Section 303 of the Act requires each state to generate a TMDL report for each body of water impaired by pollutants. TMDL reports identify pollutant levels and strategies to accomplish pollutant reduction goals. EPA has described TMDLs as establishing a "pollutant budget" with allocations to each of the pollutant's sources. For many coastal water bodies, the main pollutant issue is excess nutrients, also termed *nutrient over-enrichment*.

A TMDL can prescribe the minimum level of dissolved oxygen (DO) available in a body of water, which is directly related to nutrient levels In 2010, 18 percent of TMDLs nationwide were related to nutrient levels including organic enrichment/oxygen depletion, noxious plants, algal growth, and ammonia.

TMDLs identify all point source and nonpoint source pollutants within a watershed. To implement TMDLs with point sources, wasteload allocations are incorporated into their NPDES permits. NPS discharges are generally in a voluntary compliance scenario.

In Long Island Sound, the TMDL development process enabled the Connecticut Department of Energy and Environmental Protection and the New York State Department of Environmental Conservation to incorporate a 58.5 percent nitrogen reduction target into a regulatory and legal framework.

Nutrient Remediation

Mussels are examples of organisms that act as nutrient bioextractors.

Innovative solutions have been conceived to deal with nutrient pollution in aquatic systems by altering or enhancing natural processes to shift nutrient effects away from detrimental ecological impacts. Nutrient remediation is a form of environmental remediation, but concerns only biologically active nutrients such as nitrogen and phosphorus. "Remediation" refers to the removal of pollution or contaminants, generally for the protection of human health. In environmental remediation nutrient removal technologies include biofiltration, which uses living material to capture and biologically degrade pollutants. Examples include green belts, riparian areas, natural and constructed wetlands, and treatment ponds. These areas most commonly capture anthropogenic discharges such as wastewater, stormwater runoff, or sewage treatment, for land reclamation after mining, refinery activity, or land development. Biofiltration utilizes biological assimilation to capture, absorb, and eventually incorporate the pollutants (including nutrients) into living tissue. Another form of nutrient removal is bioremediation, which uses microorganisms to remove pollutants. Bioremediation can occur on its own as natural attenuation or intrinsic bioremediation or can be encouraged by the addition of fertilizers, a strategy called biostimulation.

Nutrient bioextraction is bioremediation involving cultured plants and animals. Nutrient bioextraction or bioharvesting is the practice of farming and harvesting shellfish and seaweed for the purpose of removing nitrogen and other nutrients from natural water bodies. It has been suggested that nitrogen removal by oyster reefs could generate net benefits for sources facing nitrogen emission restrictions, similar to other nutrient trading scenarios. Specifically, if oysters maintain nitrogen levels in estuaries below thresholds that would lead to the imposition of emission limits, oysters effectively save the sources the compliance costs they otherwise would incur. Several studies have shown that oysters and mussels have the capacity to dramatically impact nitrogen levels in estuaries. Additionally, studies have demonstrated seaweed's potential to improve nitrogen levels.

History of Nutrient Policy in the United States

The basic requirements for states to develop nutrient criteria and standards were mandated in the 1972 Clean Water Act. Implementing this water quality program has been a major scientific, technical and resource-intensive challenge for both EPA and the states, and development is continuing well into the 21st century.

EPA published a wastewater management regulation in 1978 to begin to address the national nitrogen pollution problem, which had been increasing for decades. In 1998, the agency published a *National Nutrient Strategy* with a focus on developing nutrient criteria.

Between 2000 and 2010 EPA published federal-level nutrient criteria for rivers/streams, lakes/reservoirs, estuaries and wetlands; and related guidance. "Ecoregional" nutrient criteria for 14 ecoregions across the U.S. were included in these publications. While states may directly adopt the EPA-published criteria, in many cases the states need to modfiy the criteria to reflect site-specific conditions. In 2004, EPA stated its expectations for numeric criteria (as opposed to less-specific narrative criteria) for total nitrogen (TN), total phosphorus (TP), chlorophyll a(chl-a), and clarity, and established "mutually-agreed upon plans" for state criteria development. In 2007, the agency stated that progress among the states on developing nutrient criteria had been uneven. EPA reiterated its expectations for numeric criteria and promised its support for state efforts to develop their own criteria.

In 2008 EPA published a progress report on state efforts to develop nutrient standards. A majority of states had not developed numeric nutrient criteria for rivers and streams; lakes and reservoirs; wetlands and estuaries (for those states that have estuaries). In the same year, EPA also established a Nutrient Innovations Task Group (NITG), composed of state and EPA experts, to monitor and evaluate the progress of reducing nutrient pollution. In 2009 the NTIG issued a report, "An Urgent Call to Action," expesssing concern that water quality continued to deteriorate nationwide due to increasing nutrient pollution, and recommending more vigorous development of nutrient standards by the states.

In 2011 EPA reiterated the need for states to fully develop their nutrient standards, noting that drinking water violations for nitrates had doubled in eight years, that half of all streams nationwide had medium to high levels of nitrogen and phosphorus, and harmful algal blooms were increasing. The agency set out a framework for states to develop priorities and watershed-level goals for reductions of nutrients.

Nutrient Trading

After the EPA had introduced watershed-based NPDES permitting in 2007, interest in nutrient removal and achieving regional TMDLs led to the development of nutrient trading schemes. Nutrient trading is a type of water quality trading, a market-based policy instrument used to improve or maintain water quality. Water quality trading arose around 2005 and is based on the fact that different pollution sources in a watershed can face very different costs to control the same pollutant. Water quality trading involves the voluntary exchange of pollution reduction credits from sources with low costs of pollution control to those with high costs of pollution control, and the same principles apply to nutrient water quality trading. The underlying principle is "polluter pays", usually linked with a regulatory requirement for participating in the trading program.

A 2013 Forest Trends report summarized water quality trading programs and found three main types of funders: beneficiaries of watershed protection, polluters compensating for their impacts and "public good payers" that may not directly benefit, but fund the pollution reduction credits on behalf of a government or NGO. As of 2013, payments were overwhelmingly initiated by public good payers like governments and NGOs.

Nutrient Source Apportionment

Nutrient source apportionment is used to estimate the nutrient load from various sectors entering water bodies, following attenuation or treatment. Agriculture is typically the principal source of nitrogen in water bodies in Europe, whereas in many countries households and industries tend to be the dominant contributors of phosphorus. Where water quality is impacted by excess nutrients, load source apportionment models can support the proportional and pragmatic management of water resources by identifying the pollution sources. There are two broad approaches to load apportionment modelling, (i) load-orientated approaches which apportion origin based on in-stream monitoring data and (ii) source-orientated approaches where amounts of diffuse, or nonpoint source pollution, emissions are calculated using models typically based on export coefficients from catchments with similar characteristics. For example, the Source Load Apportionment Model (SLAM) takes the latter approach, estimating the relative contribution of sources of nitrogen and phosphorus to surface waters in Irish catchments without in-stream monitoring data by integrating information on point discharges (urban wastewater, industry and septic tank systems), diffuse sources (pasture, arable, forestry, etc.), and catchment data, including hydrogeological characteristics.

Microbial Water Pollution

The source of microbiological pollution is often inadequately treated human sewage or runoff from animal husbandry facilities into streams or lakes; in addition, some microbial populations can increase in drinking water distribution systems. Other factors may also influence microbial levels, including: (1) wild animals are a reservoir for bacteria or protozoa that can infect humans; (2) variations in turbidity or water chemistry can affect bacterial densities; and (3) algal blooms may increase bacterial abundance.

Infectious Diseases in Great Lakes Region

Drinking-Water-related Outbreaks: The largest recorded waterborne disease outbreak in North America occurred in March and April, 1993, in Milwaukee. More than 400,000 people developed gastrointestinal symptoms after exposure to Cryptosporidiumcontaminated drinking water, and 4000 were hospitalized. In this outbreak, cryptosporodiosis is estimated to have contributed to an estimated 104 deaths. In 1994, an outbreak of cryptosporidiosis claimed 37 lives in Las Vegas, Nevada. Particularly worrisome is the fact that the public water facility had tested for Cryptosporidium and not detected its presence.

CDC researchers summarized U.S. data on waterborne disease outbreaks from 1993-4, in which 22 outbreaks of infectious disease were reported from 17 states and 1 territory. Of these, 10 were found to be caused by Cryptosporidium or Giardia, while 7 were caused by bacterial contamination and 5 were of unknown origin. Two-thirds involved well-water, while 23% were linked to surface water supplies. These data probably underestimate the incidence of waterborne diseases; Frost et al. estimate that only 10-33% of waterborne outbreaks are reported. Researchers in Ontario found that county-wide hospitalization rates for gastroenteritis among children (less than 18 years of age) ranged from 78.2 per 100,000 per year (Middlesex County) to 1137.6 per 100,000 per year. (Renfrew County), with an overall average of 411.1 per 100,000 per year.

Recreational Water Use-related outbreaks: The CDC7 also reports 14 infectious disease outbreaks associated with recreational water exposure, of which 10 were caused by Cryptosporidium and Giardia and 4 were caused by bacterial contamination. The four bacterial disease outbreaks were all associated with swimming in lakes, and 3 of the 4 Giardia-caused outbreaks were related to swimming in lakes or rivers. However, 5 of the 6 Cryptosporidium cases were linked to swimming in motel or community swimming pools. In addition to these outbreaks, there was one case of fatal ameobic meningoencephalitis reported in a child who had been swimming in both the Rio Grande River and a wastewater holding tank.

The USEPA has collected data on beach closings from the U.S. counties bordering the Great Lakes, most of which were due to microbial contamination. There are 582 beaches and approximately half of them are monitored for water quality. Of those monitored, 51 (18%) were closed at least once in 1993, and 80 (29%) were closed in 1994. The total number of beaches closed during a season has ranged from 16 to 80 during the period 1980-1994, and there is no particular trend in the data.

A randomized trial of the effects of seabathing on incidence of gastroenteritis and non-enteric diseases in adults was conducted in 4 resorts along the coast of the U.K. Over a 4-year period, 1216 adults were assigned to swim in coastal waters or to stay out of the water. Crude rates of gastroenteritis were significantly higher (p=0.01) in the seabathing group (14.8%) then the non-seabathing group (9.7%), and an apparent threshold level was found at 33 fecal streptococci/100 ml. Bathers had an excess risk of respiratory illness, compared to nonbathers, only when fecal streptococci exceeded 60 microorganisms per 100 ml (OR=3.92, 95% CI 1.59-9.49, p=0.0014). Risk of ear infections among bathers did not increase significantly until fecal coliform levels exceeded approximately 100 organisms/100 ml (p<0.05). When data from all study locations were grouped, an increased risk of eye infections was found in bathers (OR 2.06, 95% CI 1.01-4.25). No statistically significant trends were seen in skin ailments. The authors conclude that standards and guidelines would be more appropriately based on fecal streptococci than coliform bacteria.

Environmental Exposure Surveys

Data from a survey of raw and treated water samples from 72 municipalities in Canada indicate that Giardia cysts are commonly found in raw water samples (21%), treated water samples (18.2%) and sewage samples (73%). However, Cryptosporidium oocysts were only found in 3.5% of treated water samples and 6.1% of raw sewage samples. The authors recommend an "action level" of 3 to 5 Giardia cysts per 100 liters drinking water for public health protection.

A study in two British Columbia watersheds indicates that cattle ranches may serve as a source of protozoan contamination. Levels of Giardia cysts and Cryptosporidium oocysts were higher in samples taken downstream of cattle ranches, though the concentrations were lower in the watershed where cattle were commonly penned away from streams or lakes. In a survey conducted in the Yukon, Canada, where there have been no reported outbreaks of waterborne diseases, 32% of pristine water samples were found to be contaminated with Giardia cysts, but no Cryptosporidium oocysts were found in the samples. Both cysts and oocysts were commonly found in treated and untreated sewage.

Juranek cited findings of Cryptosporidium oocysts in over 65% of rivers and lakes tested throughout the U.S. Results of municipal water system sample analysis indicated that oocysts were present in tap water in over 27% of the communities evaluated. Paradoxically, in most outbreaks involving community drinking water supplies, the treatment facilities have been found to be in compliance with federal and state drinking water standards (the U.S. Safe Drinking Water Act standard requires that no more than 5% of drinking water samples collected in a given month can be positive for total coliform bacteria). CDC researchers found "compliance with EPA's water-treatment standards (e.g. for turbidity and coliform counts) did not adequately protect against waterborne cryptosporidiosis." As a result, in May 1996, the U.S. EPA established a requirement that, beginning in 1997, public water supply facilities serving more than 100,000 monitor for Cryptosporidium oocysts.

Other Issues

Antibiotic Resistance: According to WHO, "two important human pathogens of animal origin, E. coli and Salmonellae, are today highly resistant to antibiotics in both industrialized and developing nations." Key components contributing to the rise in antimicrobial resistance are inappropriate uses of antibiotics and their overuse in animal husbandry operations. Uncontrolled or inappropriate use of antibiotic drugs in both industrialized and developing countries has contributed to the rapid development of resistance in pathogenic microbes. In addition, "enormous amounts" of antibiotics are used in the production of animal food; the WHO estimates that "more than half of the total production of antimicrobials is used in farm animals, either for disease prevention or growth production." Drug resistant bacteria in foods can be direct causes of disease in humans, or the resistance can be transferred to human pathogens.

Global Climate Change: Some researchers have argued that one cause of the resurgence of infectious diseases, particularly in developed countries, is global climate change. The proposed mechanisms involve environmental changes or increased levels of pollutants that can bring disease-bearing wildlife into contact with humans, or cause algal blooms that may harbor or amplify microbial agents. As an example, Epstein cites evidence associating increased growth of algae or plankton, which can harbor microbial organisms or spores, with expanding ranges of cholera bacteria.

Thermal Pollution

Thermal pollution is the degradation of water quality by any process that changes ambient water temperature. A common cause of thermal pollution is the use of water as a coolant by power plants and industrial manufacturers. Other causes of thermal pollution include soil erosion. This will elevate water and expose it to sunlight. When water used as a coolant is returned to the natural environment at a higher temperature, the sudden change in temperature decreases oxygen supply and affects ecosystem composition. Fish and other organisms adapted to particular temperature range can be killed by an abrupt change in water temperature (either a rapid increase or decrease) known as "Thermal shock."

Urban runoff—stormwater discharged to surface waters from roads and parking lots—can also be a source of elevated water temperatures.

The Brayton Point Power Station in Massachusetts discharged heated water to Mount Hope Bay. The plant was shut down in June 2017.

Ecological Effects

Potrero Generating Station discharged heated water into San Francisco Bay. The plant was closed in 2011.

Warm Water Effects

Elevated temperature typically decreases the level of dissolved oxygen and of water, as gases are less soluble in hotter liquids. This can harm aquatic animals such as fish, amphibians and other aquatic organisms. Thermal pollution may also increase the metabolic rate of aquatic animals, as enzyme activity, resulting in these organisms consuming more food in a shorter time than if their environment were not changed. An increased metabolic rate may result in fewer resources; the more adapted organisms moving in may have an advantage over organisms that are not used to the warmer temperature. As a result, food chains of the old and new environments may be compromised. Some fish species will avoid stream segments or coastal areas adjacent to a thermal discharge. Biodiversity can be decreased as a result.

High temperature limits oxygen dispersion into deeper waters, contributing to anaerobic conditions. This can lead to increased bacteria levels when there is ample food supply. Many aquatic species will fail to reproduce at elevated temperatures.

Primary producers (e.g. plants, cyanobacteria) are affected by warm water because higher water temperature increases plant growth rates, resulting in a shorter lifespan and species overpopulation. The increased temperature can also change the balance of microbial growth, including the rate of algae blooms which reduce dissolved oxygen concentrations.

Temperature changes of even one to two degrees Celsius can cause significant changes in organism metabolism and other adverse cellular biology effects. Principal adverse changes can include rendering cell walls less permeable to necessary osmosis, coagulation of cell proteins, and alteration of enzyme metabolism. These cellular level effects can adversely affect mortality and reproduction.

A large increase in temperature can lead to the denaturing of life-supporting enzymes by breaking down hydrogen- and disulphide bonds within the quaternary structure of the enzymes. Decreased enzyme activity in aquatic organisms can cause problems such as the inability to break down lipids, which leads to malnutrition. Increased water temperature can also increase the solubility and kinetics of metals, which can increase the uptake of heavy metals by aquatic organisms. This can lead to toxic outcomes for these species, as well as build up of heavy metals in higher trophic levels in the food chain, increasing human exposures via dietary ingestion.

In limited cases, warm water has little deleterious effect and may even lead to improved function of the receiving aquatic ecosystem. This phenomenon is seen especially in seasonal waters and is known as thermal enrichment. An extreme case is derived from the aggregational habits of the manatee, which often uses power plant discharge sites during winter. Projections suggest that manatee populations would decline upon the removal of these discharges.

Cold Water

Releases of unnaturally cold water from reservoirs can dramatically change the fish and macroinvertebrate fauna of rivers, and reduce river productivity. In Australia, where many rivers have warmer temperature regimes, native fish species have been eliminated, and macroinvertebrate fauna have been drastically altered. This may be mitigated by designing the dam to release warmer surface waters instead of the colder water at the bottom of the reservoir.

Thermal Shock

When a power plant first opens or shuts down for repair or other causes, fish and other organisms adapted to particular temperature range can be killed by the abrupt change in water temperature, either an increase or decrease, known as "thermal shock".

Sources and Control of Thermal Pollution

Cooling tower at Gustav Knepper Power Station, Dortmund, Germany.

Industrial Wastewater

In the United States, about 75 to 82 percent of thermal pollution is generated by power plants. The remainder is from industrial sources such as petroleum refineries, pulp and paper mills, chemical plants, steel mills and smelters. Heated water from these sources may be controlled with:

- Cooling ponds, man-made bodies of water designed for cooling by evaporation, convection, and radiation.

- Cooling towers, which transfer waste heat to the atmosphere through evaporation and/or heat transfer.

- Cogeneration, a process where waste heat is recycled for domestic and/or industrial heating purposes.

Some facilities use once-through cooling (OTC) systems which do not reduce temperature as effectively as the above systems. For example, the Potrero Generating Station in San Francisco (closed in 2011), used OTC and discharged water to San Francisco Bay approximately 10 °C (20 °F) above the ambient bay temperature.

A bioretention cell for treating urban runoff in California.

Urban Runoff

During warm weather, urban runoff can have significant thermal impacts on small streams, as storm water passes over hot parking lots, roads and sidewalks. Storm water management facilities that absorb runoff or direct it into groundwater, such as bioretention systems and infiltration basins, can reduce these thermal effects. These related systems for managing runoff are components of an expanding urban design approach commonly called green infrastructure.

Retention basins (stormwater ponds) tend to be less effective at reducing runoff temperature, as the water may be heated by the sun before being discharged to a receiving stream.

Abandoned Mine Drainage

Abandoned mine drainage (also known as AMD) is a form of water pollution involving water that has been polluted by contact with mines, typically coal mines. Although it is sometimes called "acid mine drainage", not all abandoned mine drainage is acidic.

A creek affected by abandoned mine drainage.

Types

The most common form of abandoned mine drainage is acid mine drainage, which is highly acidic water coming from mines. Abandoned mine drainage can become acidic when it is exposed to oxygen and sulfur-containing minerals such as pyrite.

Another form of abandoned mine drainage is alkaline mine drainage. This typically occurs in the presence of minerals such as calcite, limestone, or dolomite. The third form of abandoned mine drainage is metal mine drainage, which occurs when large amounts of metals such as lead contaminate mine water.

Geochemistry

The precise chemistry of abandoned mine drainage discharges typically varies. Abandoned mine drainage typically originates in surface mines, deep mines, and "gob piles".

Abandoned mine drainage typically has high concentrations of metals and total dissolved solids. Iron is the most common metal in abandoned mine drainage, but aluminum and manganese occur as well. It may also have a high water temperature and an altered pH, though the characteristics of abandoned mine drainage depend heavily on the area's geochemistry. Other signs of abandoned mine drainage include high sulfate levels and siltation. Acid mine drainage has a pH of less than 7, while alkaline mine drainage has a pH of greater than 7.

The concentrations of metals in abandoned mine drainage can range from several to several thousand parts per million. In Pennsylvania's Coal Region, it has an iron concentration of less than 100 milligrams per liter (0.100 oz/cu ft) and a pH of close to 7.

Abandoned mine drainage can cause affected streams to take on a bright orange color.

Environmental Effects

Abandoned mine drainage affects streams worldwide. A 2017 United Nations Environmental Programme (UNEP) report documented "widespread destruction" from mining waste that is released into the environment as a result of dam failures.

Mine drainage is prevalent in the United States state of Pennsylvania, as well as several other states that have historically had large mining industries. In Pennsylvania, nearly 2,500 miles (4,000 km) of streams have been affected by abandoned mine drainage, and more than 7,500 miles (12,100 km) have been affected in the Appalachian Mountains, and more than 10,000 miles (16,000 km) are affected in Pennsylvania and West Virginia.

In one watershed affected by abandoned mine drainage, the value of homes within 200 feet (61 m) of an affected stream decreased by $2500 per 1 acre (0.40 ha), as of 2009.

Algae sometimes coat streams that are affected by abandoned mine drainage, although aquatic plants have difficulty surviving in such streams. The gills of fish are also harmed by the metals in abandoned mine drainage. The eggs of macroinvertebrates and fish are also smothered by the precipitate. Coldwater fish such as trout are especially harmed by abandoned mine drainage.

Abandoned mine drainage also affects the plants and animals in the area surrounding the mine. Mine drainage pollution has been found in sheep that fed near abandoned mines.

Remedies and Applications

There are several methods by which abandoned mine drainage can be remedied: passive treatment, active treatment, and land reclamation. Trompes may also be useful in treating abandoned mine drainage.

The natural gas industry has suggested using abandoned mine drainage water in hydraulic fracturing. Proponents of this idea say that it would remove toxic water from the environment while reducing the natural gas industry's need to use water from streams and rivers. However, opponents have said that it would "make a dirty process even dirtier". This has already been done with streams such as an unnamed tributary to Johnson Creek. Abandoned mine drainage solids have shown potential in capturing mercury emissions from coal-fired power plants.

The iron oxide generated by treating abandoned mine drainage can be used as a glaze for pottery.

Arsenic Contamination of Groundwater

Arsenic contamination of groundwater is a form of groundwater pollution which is often due to naturally occurring high concentrations of arsenic in deeper levels of groundwater. It is a high-profile problem due to the use of deep tubewells for water supply in the Ganges Delta, causing serious arsenic poisoning to large numbers of people. A 2007 study found that over 137 million people in more than 70 countries are probably affected by arsenic poisoning of drinking water. The problem became serious health concern after mass poisoning of water in Bangladesh. Arsenic contamination of ground water is found in many countries throughout the world, including the US.

Approximately 20 major incidents of groundwater floarsenic contamination have been reported. Of these, four major incidents occurred in Asia, in Thailand, Taiwan, and Mainland China. Locations of potentially hazardous wells have been mapped in China.

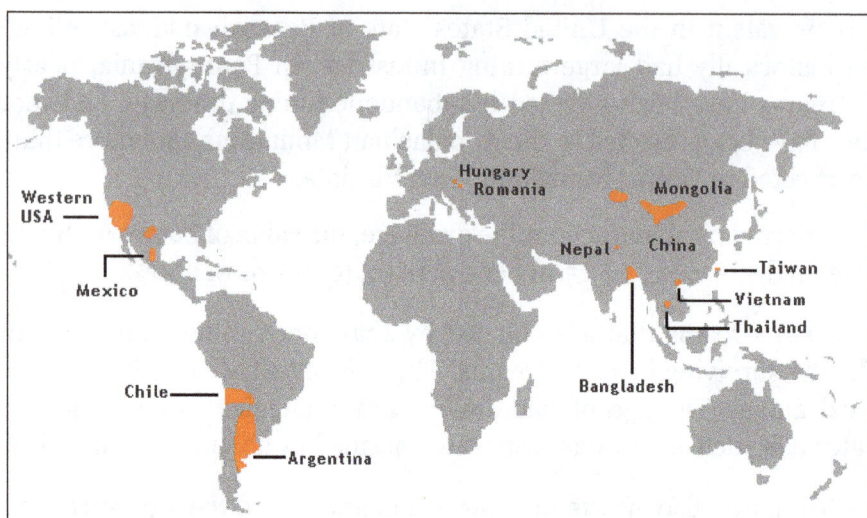

Groundwater arsenic contamination areas.

Speciation of Arsenic Compounds in Water

Arsenic contaminated water typically contains arsenous acid and arsenic acid or their derivatives. Their names as "acids" is a formality; these species are not aggressive acids but are merely the soluble forms of arsenic near neutral pH. These compounds are extracted from the underlying rocks that surround the aquifer. Arsenic acid tends to exist as the ions $[HAsO_4]^{2-}$ and $[H_2AsO_4]^-$ in neutral water, whereas arsenous acid is not ionized.

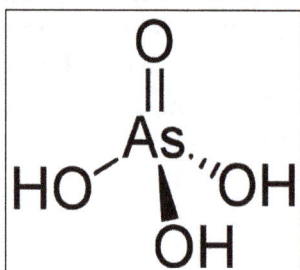

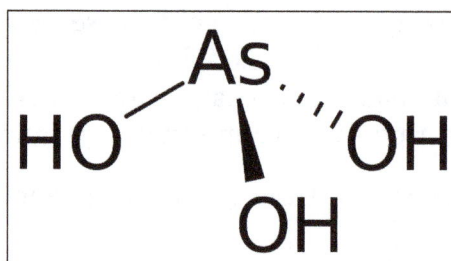

Arsenic acid (H_3AsO_4), arsenous acid (H_3AsO_3) and their derivatives are typically encountered in arsenic contaminated ground water.

Water Purification Solutions

Access to clean drinking water is fraught with political, socio-economic, and cultural inequities. In practice, many water treatment strategies tend to be temporary fixes to a larger problem, often prolonging the social issues while treating the scientific ones. Scientific studies have shown that interdisciplinary approaches to water purification are especially important to consider, and long-lasting improvements involve larger perspectives than strict scientific approaches.

Small-scale Water Treatment

A review of methods to remove arsenic from groundwater in Pakistan summarizes the most technically viable inexpensive methods. Most small-scale treatments focus on water after it has left the distribution site, and are thus more focused on quick, temporary fixes.

A simpler and less expensive form of arsenic removal is known as the Sono arsenic filter, using three pitchers containing cast iron turnings and sand in the first pitcher and wood activated carbon and sand in the second. Plastic buckets can also be used as filter containers. It is claimed that thousands of these systems are in use and can last for years while avoiding the toxic waste disposal problem inherent to conventional arsenic removal plants. Although novel, this filter has not been certified by any sanitary standards such as NSF, ANSI, WQA and does not avoid toxic waste disposal similar to any other iron removal process.

In the United States small "under the sink" units have been used to remove arsenic from drinking water. This option is called "point of use" treatment. The most common types of domestic treatment use the technologies of adsorption (using media such as Bayoxide E33, GFH, activated alumina or titanium dioxide) or reverse osmosis. Ion exchange and activated alumina have been considered but not commonly used.

Chaff-based filters have been reported to reduce the arsenic content of water to 3 μg/L (3 ppb). This is especially important in areas where the potable water is provided by filtering the water extracted from the underground aquifer.

In iron electrocoagulation (Fe-EC), iron is dissolved nonstop using electricity, and the resulting ferric hydroxides, oxyhydroxides, and oxides form an absorbent readily attracted to arsenic. Current density, the amount of charge delivered per liter of water, of the process is often manipulated in order to achieve maximum arsenic depletion. This treatment strategy has primarily been used in Bangladesh, and has proven to be largely successful. In fact, using iron electrocoagulation to remove arsenic in water proved to be the most effective treatment option.

Large-scale Water Treatment

In some places, such as the United States, all the water supplied to residences by utilities must meet primary (health-based) drinking water standards. Regulations may need large-scale treatment systems to remove arsenic from the water supply. The effectiveness of any method depends on the chemical makeup of a particular water supply. The aqueous chemistry of arsenic is complex, and may affect the removal rate that can be achieved by a particular process.

Some large utilities with multiple water supply wells could shut down those wells with high arsenic concentrations, and produce only from wells or surface water sources that meet the arsenic standard. Other utilities, however, especially small utilities with only a few wells, may have no available water supply that meets the arsenic standard.

Coagulation/filtration (also known as flocculation) removes arsenic by coprecipitation and adsorption using iron coagulants. Coagulation/filtration using alum is already used by some utilities to remove suspended solids and may be adjusted to remove arsenic.

Iron oxide adsorption filters the water through a granular medium containing ferric oxide. Ferric oxide has a high affinity for adsorbing dissolved metals such as arsenic. The iron oxide medium eventually becomes saturated, and must be replaced. The sludge disposal is a problem here too.

Activated alumina is an adsorbent that effectively removes arsenic. Activated alumina columns connected to shallow tube wells in India and Bangladesh have removed both As(III) and As(V) from groundwater for decades. Long-term column performance has been possible through the efforts of community-elected water committees that collect a local water tax for funding operations and maintenance. It has also been used to remove undesirably high concentrations of fluoride.

Ion exchange has long been used as a water softening process, although usually on a single-home basis. Traditional anion exchange resins are effective in removing As(V), but not As(III), or arsenic trioxide, which doesn't have a net charge. Effective long-term ion exchange removal of arsenic requires a trained operator to maintain the column.

Both reverse osmosis and electrodialysis (also called electrodialysis reversal) can remove arsenic with a net ionic charge. (Note that arsenic oxide, As_2O_3, is a common form of arsenic in groundwater that is soluble, but has no net charge.) Some utilities presently use one of these methods to reduce total dissolved solids and therefore improve taste. A problem with both methods is the production of high-salinity waste water, called brine, or concentrate, which then must be disposed of.

Subterranean Arsenic Removal (SAR) Technology

In subterranean arsenic removal, aerated groundwater is recharged back into the aquifer to create an oxidation zone which can trap iron and arsenic on the soil particles through adsorption process. The oxidation zone created by aerated water boosts the activity of the arsenic-oxidizing microorganisms which can oxidize arsenic from +3 to +5 state SAR Technology. No chemicals are used and almost no sludge is produced during operational stage since iron and arsenic compounds are rendered inactive in the aquifer itself. Thus toxic waste disposal and the risk of its future mobilization is prevented. Also, it has very long operational life, similar to the long lasting tube wells drawing water from the shallow aquifers.

Six such SAR plants, funded by the World Bank and constructed by Ramakrishna Vivekananda Mission, Barrackpore & Queen's University Belfast, UK are operating in West Bengal. Each plant has been delivering more than 3,000 litres of arsenic and iron-free water daily to the rural community. The first community water treatment plant based on SAR technology was set up at Kashimpore near Kolkata in 2004 by a team of European and Indian engineers led by Bhaskar Sen Gupta of Queen's University Belfast for TiPOT.

SAR technology had been awarded Dhirubhai Ambani Award, 2010 from IChemE UK for Chemical Innovation. Again, SAR was the winner of the St. Andrews Award for Environment, 2010. The SAR Project was selected by the Blacksmith Institute – New York & Green Cross- Switzerland as one of the "12 Cases of Cleanup & Success" in the World's Worst Polluted Places Report 2009.

Currently, large scale SAR plants are being installed in US, Malaysia, Cambodia, and Vietnam.

Dietary Intake

Researchers from Bangladesh and the United Kingdom have claimed that dietary intake of arsenic adds a significant amount to total intake where contaminated water is used for irrigation.

References

- "Groundwater pollution is much more difficult to abate than surface pollution". Www.coursehero.com. Retrieved 2019-08-06

- Types-of-water-pollution: ibanplastic.com, Retrieved 22 March, 2019

- Gutierrez T, Berry D, Teske A, Aitken MD (2016). "Enrichment of Fusobacteria in Sea Surface Oil Slicks from the Deepwater Horizon Oil Spill". Microorganisms. 4 (3): 24. Doi:10.3390/microorganisms4030024. PMC 5039584. PMID 27681918

- Hamblin, Jacob Darwin (2008). Poison in the Well: Radioactive Waste in the Oceans at the Dawn of the Nuclear Age. Rutgers University Press. ISBN 978-0-8135-4220-1

- What-is-surface-water-pollution-sources-effects: aosts.com, Retrieved 23 April, 2019

- Gwinnett, Claire (2019-02-17). "The major source of plastic pollution you've probably never heard of". Fast Company. Retrieved 2019-02-18

- Types-of-water-pollution: ibanplastic.com, Retrieved 24 May, 2019

- Khan, M. Nasir and Mohammad, F. (2014) "Eutrophication of Lakes" in A. A. Ansari, S. S. Gill (eds.), Eutrophication: Challenges and Solutions; Volume II of Eutrophication: Causes, Consequences and Control, Springer Science+Business Media Dordrecht. Doi:10.1007/978-94-007-7814-6_5. ISBN 978-94-007-7814-6

- Chemical-water-pollution: water-pollution.org.uk, Retrieved 25 June, 2019

3

Sources of Water Pollution

The main sources of water pollution are domestic effluents and sewage, industrial effluents, agricultural effluents, radioactive wastes, thermal pollution and oil pollution. The topics elaborated in this chapter will help in gaining a better perspective about these sources of water pollution.

Main Sources of Water Pollution

Some of the important sources of water pollution are: (i) Domestic effluents and sewage, (ii) Industrial effluents, (iii) Agricultural effluents, (iv) Radioactive wastes, (v) Thermal pollution, and (vi) Oil pollution.

Domestic Effluents and Sewage

Man, for his various domestic purposes such as drinking, cooking, bathing, cleaning, cooling, etc., uses on an average 135 litres of water per day. About 70 to 80 per cent of this is discharged and drained out, which through municipal drains poured into, in many cases, a river, tank or lake

This water is known as domestic waste water and, when other waste material such as paper, plastic, detergents, cloth, etc., is mixed in it; it becomes municipal waste or sewage.

The domestic waste water and sewage is the main source of the water pollution. This is the inevitable and unfortunate fallout of urbanisation. This organic waste depletes the oxygen from water and upsets the natural balance of the aquatic ecosystem.

Municipal sewage is considered to be the main pollutant of water. Most of the sewage receives no treatment before discharge, especially in developing countries like India. In Delhi alone, 120 crore litres of water is consumed per day, out of which 96 crore litres is drained into the Yamuna river through 17 big drains. In the same manner, all the 47 towns located on the bank of river Ganga drain their sewage into it.

With the growth of population, the quantity of waste water is also increasing in addition to the production of large quantities of sewage. Sewage contains decomposable organic matter and exert an oxygen demand on the receiving waters.

The common organic materials found in sewage are soaps, synthetic detergents, fatty acids, and proteinaceous matters such as amines, amino acids, amides and amino sugars.

Besides, it also contains numerous micro-organisms in the form of pathogenic bacteria and viruses derived from human faces. Untreated waste water is often the carrier of viruses and bacteria and, with poor household sanitation practices, has been linked to high infant mortality rates in developing countries.

Even where most sewage is treated, as in the developed world, recent measurements of fecal coli forms in some countries indicate increasing pollution. Sewage supports the growth of other forms of life that consume oxygen; it is measured in terms of Biochemical Oxygen Demand (BOD). It is the lack of oxygen that kills fish and other aquatic life.

In recent years, there has been considerable growth in the use of detergents, which causes severe water pollution. Many modern detergents contain phosphates, which are an essential component of agricultural fertilisers. When phosphate detergents are discharged into waterways, they supply a needed nutrient and promote rapid growth of algae.

This enrichment process is known as 'eutrophication'. In many areas of the world, aquatic weeds have multiplied explosively. They have interfered with fishing, navigation, irrigation and even production of hydroelectricity. In developing countries, human population and settlements are growing fast, often faster than waste water treatment facilities can be provided.

Thus, much of the untreated waste water and sewage is discharged into rivers and other water bodies, making the water unsuitable for drinking.

Industrial Effluents

Industrial activities generate a wide variety of waste products, which are normally discharged into water courses. Major contributors are the pulp and paper, chemicals, petrochemicals and refining, metal working, food processing, textile, distillery, etc. The wastes, broadly categorised as heavy metals or synthetic organic compounds, reach bodies of water either through direct discharge or by leaching from waste dumps.

In developed countries of the world, many industrial discharges are strictly controlled. Yet, water pollution continues from accumulations of wastes discharged over the past 100 years. But, in developing countries, industrial discharges are largely uncontrolled and thus, a major cause of water pollution.

All the Indian rivers have been polluted by industrial effluents. The 'holy' river Ganga has become a highly polluted river today due to various types of industrial discharges. Along the Ganga, several chemical, textile, tanning, pulp and paper, petrochemical, rubber, fertiliser and other industries are located and all of them discharge their waste water and other effluents, directly or indirectly, into the river, resulting in the pollution to such an extent that even the Ganga Action Plan, to control water pollution, has failed miserably.

From Delhi industrial area alone, more than 8 lakh tonnes of industrial waste is discharged into the river Yamuna.

Damodar river of Bihar is a highly polluted river due to industrial wastes discharged from Bokaro, Rourkela, Indian Iron and Steel Company (IISCO), Bengal Paper Mills, Sindhri Fertiliser Factory, etc.

A study reveals that from Durgapur Steel Plant 1,800 cu/m washed coal has been discharged into the river. Similarly, from IISCO 15,000 cu/m and from Bengal Paper Mills 12,000 cu/m industrial waste is discharged per day into river water.

Likewise, about 10 to 15 tonnes of sulfuric acid from Sindri and 5 to 10 tonnes toxic chemicals are discharged into the river water. The story of the Hooghly river in West Bengal is more or less same. Its water has been polluted to such an extent that even fish fertilisation has become difficult.

The paper mills located along the river discharge about 11.4 tonnes liquid wastes into the river. Almost all other rivers of India have the same fate. Chambal, Narmada, Kaveri, Godavari, Mahanadi and all other small rivers have been polluted and if this is not stopped, it will result in greater water pollution.

The nature and effects of the pollutants from the effluents of paper and pulp, textile, food processing, chemicals, metal and petroleum industries are as follows:

1. Effluents from paper and pulp industries include wood chips, bits of bark, cellulose fibres and dissolved lignin, in addition to a mixture of chemicals. All these produce a sludge which blankets fish spawning grounds and destroys certain types of aquatic life.

    The effluents contain chlorine, sulfur dioxide, methyl mercaptan, etc., which are considered to be highly poisonous to fish.

2. Effluents from textile industries are alkaline in nature and have a higher demand for oxygen.

3. Food processing industries include dairies, breweries, distilleries, meat-packing, etc., where the waste products include fats, proteins and organic wastes.

    These industries discharge wastes containing nitrogen, sugar, proteins, etc. This waste contains a higher BOD and is therefore responsible for water pollution.

4. Chemical industries include acid manufacturing, alkali manufacturing, fertiliser, pesticides and several other industries. The effluents from these industries contain acids which have corrosive effects. The effluents from fertiliser industries contain phosphorus, fluorine, silica, and a large amount of suspended solids.

5. Metal industries usually discharge effluents containing copper, lead, chromium, cadmium, zinc, etc., which are toxic to man as well to aquatic life. These wastes also contain acids, oils, greases and cleansing agents.

6. Petroleum industries include oil refineries and petrochemical plants. The effluents include hydrocarbons, phenolic compounds and other organic and inorganic sulphur compounds.

7. Other industries, which pollute water, are tanneries, soaps and detergent industries, glass, electroplating, bleaching, atomic plants, explosive factories, etc.

8. Mining operations can result in metals leaching into the acidic effluents, thus adding to the metal load in rivers, lakes and groundwater. Discharge of mercury from gold mining activities has polluted some streams in Brazil and Ecuador and created serious health problems.

With reference to water pollution through mercury, mention of Minimata Gulf incident must be made. In 1950, near the Japanese coast, in Minimata Gulf, fishermen suffered from blindness, weakness, mental illness, paralysis, etc.

It was found that effluents discharged from a plastic factory contained mercury, which entered the fish and by eating those fish, all the fishermen suffered from effects of mercury poisoning. Thus water pollution through industrial effluents has become a major environmental problem today and needs measures to control it.

Agricultural Effluents

Agricultural water pollution is caused by fertilisers, insecticides and pesticides, farm animal wastes and sediments. In recent years, use of chemical fertilisers has increased manifold. The green revolution in India is a reflection of the increased use of fertilisers. The chemicals used in fertilisers enter the groundwater by leaching and the surface waters by run-off.

The nitrates, when mixed with water, may cause methemoglobinemia in infants. Incidences of nitrate poisoning are also seen in livestock. The plant nutrients, nitrogen and phosphorus are reported to stimulate the growth of algae and other aquatic plants.

The use of various types of pesticides and insecticides in agriculture is also one of the causes of water pollution. Their presence in water is highly toxic to both man and animals, because these entire have a high persistence capacity, i.e., their residues remain for long periods.

The farm animal wastes often pose serious problems of odour and water pollution. These wastes also contain pathogenic organisms which get transmitted to humans. Sediments of soil and mineral particles washed out from fields also cause water pollution. They fill stream channels and reservoirs and reduce the sunlight available to aquatic plants.

Radioactive Wastes

Radioactive elements, such as uranium and radium, possess highly unstable atomic nuclei. This disintegration results in radiation emission which may be highly injurious. During nuclear tests, radioactive dust may encircle the globe at altitudes of 3,000 metres or more, which often comes down to the earth as rain.

Eventually, some of the radioactive material, such as Strontium 90 (which can cause bone cancer), percolates down through the soil into groundwater reservoirs or is carried out into streams and rivers.

In both cases, public water supplies may be contaminated. The construction of more nuclear reactors and the increasing use of radioactive materials in medical research represent other potential contamination sources.

Thermal Pollution

Most of the thermal and electric power plants also discharge considerable quantities (about 66%) of hot effluent/water into nearby streams or rivers. This has resulted in thermal pollution of our water courses. Thermal pollution is undesirable for several reasons. Warm water does not have the same oxygen holding capacity as cold water.

Therefore, fishes like black bass, trout and walleyes, etc., which require a minimal oxygen concentration of about 4 ppm, would either have to emigrate from the polluted area or die in large numbers. When the temperature of the receiving water is raised, the dissolved oxygen level decreases and the demand for oxygen increases, hence anaerobic conditions will set in resulting in the release of foul gases.

Thermal pollution is considered hazardous for the whole aquatic ecosystem. Several industries have installed cooling towers, where the heated water is cooled. But even so, thermal pollution has become a serious problem for water bodies located near thermal plants

Oil Pollution

The spread of oil in the sea has become a common feature nowadays. Oil is transported across oceans through tankers and either due to some accident or leakage oil spills onto the water and causes the degradation of aquatic and marine environment. Between 1968 and 1983, there were more than 500 tanker accidents that involved oil spills. Altogether, more than one million tonnes of oil was released. A dramatic incident was that of the tanker Torrey Canyon, when it struck off the southern tip of the British Isles in March 1967. The Torry Canyon was the largest oil spill up to that time. The pollution caused widespread destruction of many- forms of marine life despite strenuous efforts to clean up the spill. Similarly, on March 16, 1978, the oil tanker Amoco Cadiz lost its steering off the coast of Brittany in France and the total spilling of oil was 1.6 million barrels.

Such accidents have become very common due to technical problems or heavy marine traffic. During the 1991 Gulf War, there was heavy bombing on oil tanks, which resulted in the spilling of oil.

The impact of this oil spill on the marine ecosystem in this area has not yet been remedied. Offshore drilling operations also contribute their share of oil to the sea. The total quantity of oil that finds its way into sea each year is very large. It has been estimated that about one million tonnes of oil spills into the ocean each year from tankers and oil drilling operations.

Point Source of Water Pollution

Point source water pollution involves pollutants that are discharged from any identifiable, singular source, such as a pipe, ditch, drain, channel, tunnel, conduit, well, container, or vessel. By contrast, nonpoint source water pollution is caused by broadly distributed and disconnected sources of pollution, such as rain and snowmelt runoff, spills, leaks, and sediment erosion.

The Clean Water Act makes it unlawful to discharge any pollutant from a point source into navigable waters without a permit. Point source water pollution is regulated under the National Pollutant Discharge Elimination System (NPDES) permit program. Regulated pollutants from point sources include wastes, soils, rocks, chemicals, bacteria, suspended solids, heavy metals, pesticides, and more.

The Clean Water Act established water quality standards for surface waters, programs regulating the discharge of industrial and municipal wastewater, regulation of nonpoint source water pollution (pollution with no single, identifiable source) as well as point source pollution, and federal funding for municipal sewage treatment plants.

The 1972 amendments to the Clean Water Act created the National Pollutant Discharge Elimination System (NPDES), requiring authorization (from the relevant state or federal agency) to discharge pollutants from a point source into navigable waters. NPDES permits contain technology-based limits on discharges of pollutants from point sources known as technology-based effluent limits. These limits are based on how well a treatment method can limit pollutant levels to a certain concentration before the pollutants are discharged into a body of water. In addition to limiting pollutant discharges using the NPDES system, the act requires enforceable water quality standards to maintain overall water quality. Standards are set for bodies of water based on the water's designated use, such as industrial water supply, swimming, fishing, agricultural irrigation, and more. The act also requires that standards include the maximum concentrations of various pollutants that would inhibit a waterway's designated use. States establish water quality standards for waterways within their borders, though the EPA may disapprove and replace state standards with its own if they do not meet the act's minimum requirements.

State governments monitor waterways to ensure that bodies of water meet standards. For waters that do not meet quality standards, states use two additional anti-pollution methods to ensure impaired water bodies ultimately meet standards. First, states will set total maximum daily loads (TMDLs), which are the maximum allowable amounts of a pollutant that may be discharged into impaired bodies of water. TMDLs are set with the goal of reducing pollution from point sources so a body of water can meet quality standards. Second, states will divide the maximum allowable amount of a pollutant discharge into an impaired water among various pollution sources.

Types of Point Sources

Examples of point sources include sewage treatment plants; oil refineries; paper and pulp mills; chemical, automobile, and electronics manufacturers; and factories. Regulated pollutants from point sources include wastes, soils, rocks, chemicals, bacteria, suspended solids, heavy metals, pesticides, and more.

The images below show examples of point sources.

An industrial facility located along the Calumet River near Chicago, Illinois.

A sewage treatment facility.

A paper mill in Duluth, Minnesota.

Nonpoint Source Pollution

Nonpoint source (NPS) pollution is pollution resulting from many diffuse sources, in direct contrast to point source pollution which results from a single source. Nonpoint source pollution generally results from land runoff, precipitation, atmospheric deposition, drainage, seepage, or hydrological modification (rainfall and snowmelt) where tracing pollution back to a single source is difficult.

Non-point source water pollution affects a water body from sources such as polluted runoff from agricultural areas draining into a river, or wind-borne debris blowing out to sea. Non-point source air pollution affects air quality from sources such as smokestacks or car tailpipes. Although these pollutants have originated from a point source, the long-range transport ability and multiple sources of the pollutant make it a non-point source of pollution. Non-point source pollution can be contrasted with point source pollution, where discharges occur to a body of water or into the atmosphere at a single location.

NPS may derive from many different sources with no specific solution may change to rectify the problem, making it difficult to regulate. Non point source water pollution is difficult to control because it comes from the everyday activities of many different people, such as lawn fertilization, applying pesticides, road construction or building construction.

Muddy river

It is the leading cause of water pollution in the United States today, with polluted runoff from agriculture and hydromodification the primary sources.

Other significant sources of runoff include habitat modification and silviculture (forestry).

Contaminated stormwater washed off parking lots, roads and highways, and lawns (often containing fertilizers and pesticides) is called urban runoff. This runoff is often classified as a type of NPS pollution. Some people may also consider it a point source because many times it is channeled into municipal storm drain systems and discharged through pipes to nearby surface waters. However, not all urban runoff flows through storm drain systems before entering water bodies. Some may flow directly into water bodies, especially in developing and suburban areas. Also, unlike other types of point sources, such as industrial discharges, sewage treatment plants and other operations, pollution in urban runoff cannot be attributed to one activity or even group of activities. Therefore, because it is not caused by an easily identified and regulated activity, urban runoff pollution sources are also often treated as true non-point sources as municipalities work to abate them.

Principal Types

Runoff of soil and fertilizer during a rain storm.

Sediment

Sediment (loose soil) includes silt (fine particles) and suspended solids (larger particles). Sediment may enter surface waters from eroding stream banks, and from surface runoff due to improper plant cover on urban and rural land. Sediment creates turbidity (cloudiness) in water bodies, reducing the amount of light reaching lower depths, which can inhibit growth of submerged aquatic plants and consequently affect species which are dependent on them, such as fish and shellfish. High turbidity levels also inhibit drinking water purification systems.

Sediment can also be discharged from multiple different sources. Sources include construction sites (although these are point sources, which can be managed with erosion controls and sediment controls), agricultural fields, stream banks, and highly disturbed areas.

Nutrients

Nutrients mainly refers to inorganic matter from runoff, landfills, livestock operations and crop lands. The two primary nutrients of concern are phosphorus and nitrogen.

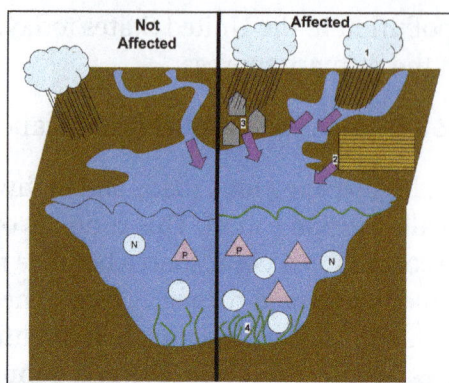

Nonpoint source pollution is caused when precipitation (1) carries pollutants from the ground such as nitrogen (N) and phosphorus (P) pollutants which come from fertilizers used on farm lands (2) or urban areas (3). These nutrients can cause eutrophication (4).

Phosphorus is a nutrient that occurs in many forms that are bioavailable. It is notoriously over-abundant in human sewage sludge. It is a main ingredient in many fertilizers used for agriculture as well as on residential and commercial properties, and may become a limiting nutrient in freshwater systems and some estuaries. Phosphorus is most often transported to water bodies via soil erosion because many forms of phosphorus tend to be adsorbed on to soil particles. Excess amounts of phosphorus in aquatic systems (particularly freshwater lakes, reservoirs, and ponds) leads to proliferation of microscopic algae called phytoplankton. The increase of organic matter supply due to the excessive growth of the phytoplankton is called eutrophication. A common symptom of eutrophication is algae blooms that can produce unsightly surface scums, shade out beneficial types of plants, produce taste-and-odor-causing compounds, and poison the water due to toxins produced by the algae. These toxins are a particular problem in systems used for drinking water because some toxins can cause human illness and removal of the toxins is difficult and expensive. Bacterial decomposition of algal blooms consumes dissolved oxygen in the water, generating hypoxia with detrimental consequences for fish and aquatic invertebrates.

Nitrogen is the other key ingredient in fertilizers, and it generally becomes a pollutant in saltwater or brackish estuarine systems where nitrogen is a limiting nutrient. Similar to phosphorus in fresh-waters, excess amounts of bioavailable nitrogen in marine systems lead to eutrophication and algae blooms. Hypoxia is an increasingly common result of eutrophication in marine systems and can impact large areas of estuaries, bays, and near shore coastal waters. Each summer, hypoxic conditions form in bottom waters where the Mississippi River enters the Gulf of Mexico. During recent summers, the aerial extent of this "dead zone" is comparable to the area of New Jersey and has major detrimental consequences for fisheries in the region.

Nitrogen is most often transported by water as nitrate (NO_3). The nitrogen is usually added to a watershed as organic-N or ammonia (NH_3), so nitrogen stays attached to the soil until oxidation converts it into nitrate. Since the nitrate is generally already incorporated into the soil, the water traveling through the soil (i.e., interflow and tile drainage) is the most likely to transport it, rather than surface runoff.

Toxic Contaminants and Chemicals

Compounds including heavy metals like lead, mercury, zinc, and cadmium, organics like polychlorinated biphenyls (PCBs) and polycyclic aromatic hydrocarbons (PAHs), fire retardants, and other substances are resistant to breakdown. These contaminants can come from a variety of sources including human sewage sludge, mining operations, vehicle emissions, fossil fuel combustion, urban runoff, industrial operations and landfills.

Toxic chemicals mainly include organic compounds and inorganic compounds. These compounds include pesticides like DDT, acids, and salts that have severe effects to the ecosystem and water-bodies. These compounds can threaten the health of both humans and aquatic species while being resistant to environmental breakdown, thus allowing them to persist in the environment. These toxic chemicals could come from croplands, nurseries, orchards, building sites, gardens, lawns and landfills.

Acids and salts mainly are inorganic pollutants from irrigated lands, mining operations, urban runoff, industrial sites and landfills.

Pathogens

Pathogens are bacteria and viruses that can be found in water and cause diseases in humans. Typically, pathogens cause disease when they are present in public drinking water supplies. Pathogens found in contaminated runoff may include:

- Cryptosporidium parvum
- Giardia lamblia
- Salmonella
- Norovirus and other viruses
- Parasitic worms (helminths)

Coliform bacteria and fecal matter may also be detected in runoff. These bacteria are a commonly used indicator of water pollution, but not an actual cause of disease.

Pathogens may contaminate runoff due to poorly managed livestock operations, faulty septic systems, improper handling of pet waste, the over application of human sewage sludge, contaminated storm sewers, and sanitary sewer overflows.

Principal Sources

Urban and Suburban Areas

Urban and suburban areas are a main sources of nonpoint source pollution due to the amount of runoff that is produced due to the large amount of paved surfaces. Paved surfaces, such as asphalt and concrete are impervious to water penetrating them. Any water that is on contact with these surfaces will run off and be absorbed by the surrounding environment. These surfaces make it easier for stormwater to carry pollutants into the surrounding soil.

Construction sites tend to have disturbed soil that is easily eroded by precipitation like rain, snow, and hail. Additionally, discarded debris on the site can be carried away by runoff waters and enter the aquatic environment.

Typically, in suburban areas, chemicals are used for lawn care. These chemicals can end up in run-off and enter the surrounding environment via storm drains in the city. Since the water in storm drains is not treated before flowing into surrounding water bodies, the chemicals enter the water directly.

Agricultural Operations

Agricultural operations account for a large percentage of all nonpoint source pollution in the United States. When large tracts of land are plowed to grow crops, it exposes and loosens soil that was once buried. This makes the exposed soil more vulnerable to erosion during rainstorms. It also can increase the amount of fertilizer and pesticides carried into nearby bodies of water.

Atmospheric Inputs

Atmospheric deposition is a source of inorganic and organic constituents because these constituents are transported from sources of air pollution to receptors on the ground. Typically, industrial facilities, like factories, emit air pollution via a smokestack. Although this is a point source, due to the distributional nature, long-range transport, and multiple sources of the pollution, it can be considered as nonpoint source in the depositional area. Atmospheric inputs that affect runoff quality may come from dry deposition between storm events and wet deposition during storm events. The effects of vehicular traffic on the wet and dry deposition that occurs on or near highways, roadways, and parking areas creates uncertainties in the magnitudes of various atmospheric sources in runoff. Existing networks that use protocols sufficient to quantify these concentrations and loads do not measure many of the constituents of interest and these networks are too sparse to provide good deposition estimates at a local scale.

Highway Runoff

Highway runoff accounts for a small but widespread percentage of all nonpoint source pollution. Harned estimated that runoff loads were composed of atmospheric fallout (9%), vehicle deposition (25%) and highway maintenance materials (67%) he also estimated that about 9 percent of these loads were reentrained in the atmosphere.

Forestry and Mining Operations

Forestry and mining operations can have significant inputs to non-point source pollution.

Forestry

Forestry operations reduce the number of trees in a given area, thus reducing the oxygen levels in that area as well. This action, coupled with the heavy machinery (harvesters, etc.) rolling over the soil increases the risk of erosion.

Mining

Active mining operations are considered point sources, however runoff from abandoned mining operations contribute to nonpoint source pollution. In strip mining operations, the top of the mountain is removed to expose the desired ore. If this area is not properly reclaimed once the mining has finished, soil erosion can occur. Additionally, there can be chemical reactions with the air and newly exposed rock to create acidic runoff. Water that seeps out of abandoned subsurface mines can also be highly acidic. This can seep into the nearest body of water and change the pH in the aquatic environment.

Marinas and Boating Activities

Chemicals used for boat maintenance, like paint, solvents, and oils find their way into water through runoff. Additionally, spilling fuels or leaking fuels directly into the water from boats contribute to nonpoint source pollution. Nutrient and bacteria levels are increased by poorly maintained sanitary waste receptacles on the boat and pump-out stations.

Control

Regulation of Nonpoint Source Pollution

Contour buffer strips used to retain soil and reduce erosion.

The definition of a nonpoint source is addressed under the Clean Water Act by the Environmental Protection Agency. However, the EPA does not provide direct terms of regulation of nonpoint sources. Instead, the agency leaves regulation of nonpoint sources up to state and local governments. For example, many states have taken the steps to implement their own management programs for places such as their coastlines, all of which have to be approved by both NOAA and the EPA. The goals of these programs and those alike are to create foundations that encourage statewide pollution reduction by growing and improving systems that already exist. Programs within these state and local governments look to Best Management Practices (BMPs) in order to accomplish their goals of finding the least costly method to reduce the greatest amount of pollution. BMPs can be implemented for both agricultural and urban runoff, and can also be either structural or nonstructural methods. Many agencies, like the EPA and USDA, have approved and provided a list of commonly used BMPs for the many different categories of nonpoint source pollution.

Clean Water Act Provisions for States

319 Grant Program for States and Territories

The 319 Grant Program for States and Territories is an amendment made to the Clean Water Act in 1987 where grant money is given to states, territories, ann tribes in order to incentivize implementation and further development in policy. This amendment requires nonpoint source (NPS) management programs to be utilized in all states. The EPA mandates updates to NPS programs state-to-state in order to effectively manage the ever-changing nature of their waters, and to ensure effective use of the 319 grant money and resources.

Coastal Zone Act Reauthorization Amendments (CZARA)

The Coastal Zone Act Reauthorization Amendments (CZARA) is another provision of the Clean Water Act that mandates regulation of nonpoint source pollution in states with coastal waters. CZARA requires states with coastlines to implement management measures to remediate water pollution, and to make sure that the product of these measures is implementation as opposed to adoption.

Urban and Suburban Areas

To control non-point source pollution, many different approaches can be undertaken in both urban and suburban areas. Buffer strips provide a barrier of grass in between impervious paving material like parking lots and roads, and the closest body of water. This allows the soil to absorb any pollution before it enters the local aquatic system. Retention ponds can be built in drainage areas to create an aquatic buffer between runoff pollution and the aquatic environment. Runoff and storm water drain into the retention pond allowing for the contaminants to settle out and become trapped in the pond. The use of porous pavement allows for rain and storm water to drain into the ground beneath the pavement, reducing the amount of runoff that drains directly into the water body. Restoration methods such as constructing wetlands are also used to slow runoff as well as absorb contamination.

Construction sites typically implement simple measures to reduce pollution and runoff. Firstly, sediment or silt fences are erected around construction sites to reduce the amount of sediment and large material draining into the nearby water body. Secondly, laying grass or straw along the border of construction sites also work to reduce nonpoint source pollution.

In areas served by single-home septic systems, local government regulations can force septic system maintenance to ensure compliance with water quality standards. In Washington (state), a novel approach was developed through a creation of a "shellfish protection district" when either a commercial or recreational shellfish bed is downgraded because of ongoing nonpoint source pollution. The shellfish protection district is a geographic area designated by a county to protect water quality and tideland resources, and provides a mechanism to generate local funds for water quality services to control nonpoint sources of pollution. At least two shellfish protection districts in south Puget Sound have instituted septic system operation and maintenance requirements with program fees tied directly to property taxes.

Agricultural Operations

To control sediment and runoff, farmers may utilize erosion controls to reduce runoff flows and retain soil on their fields. Common techniques include contour plowing, crop mulching, crop rotation, planting perennial crops or installing riparian buffers. *Conservation tillage* is a concept used to reduce runoff while planting a new crop. The farmer leaves some crop reside from the previous planting in the ground to help prevent runoff during the planting process.

Nutrients are typically applied to farmland as commercial fertilizer; animal manure; or spraying of municipal or industrial wastewater (effluent) or sludge. Nutrients may also enter runoff from crop residues, irrigation water, wildlife, and atmospheric deposition. Farmers can develop and implement nutrient management plans to reduce excess application of nutrients.

To minimize pesticide impacts, farmers may use Integrated Pest Management (IPM) techniques (which can include biological pest control) to maintain control over pests, reduce reliance on chemical pesticides, and protect water quality.

Forestry Operations

With a well-planned placement of both logging trails, also called skid trails, can reduce the amount of sediment generated. By planning the trails location as far away from the logging activity as possible as well as contouring the trails with the land, it can reduce the amount of loose sediment in the runoff. Additionally, by replanting trees on the land after logging, it provides a structure for the soil to regain stability as well as replaces the logged environment.

Marinas

Installing shut off valves on fuel pumps at a marina dock can help reduce the amount of spillover into the water. Additionally, pump-out stations that are easily accessible to boaters in a marina can provide a clean place in which to dispose of sanitary waste without dumping it directly into the water. Finally, something as simple as having trash containers around a marina can prevent larger objects entering the water.

4
Effects of Water Pollution

The effects of water pollution depend on the chemicals and concentration of water pollutants. Some of the adverse effects which can be caused by water pollution are killing of aquatic animals, disruption of food chains and their habitat, infectious diseases and destruction of ecosystems. All these effects of water pollution have been carefully analyzed in this chapter.

Many different types of industrial sites produce mercury as a byproduct while they are in operation. When this happens, the mercury can easily run into groundwater and even surface water without being detected. Mercury poisoning can be very dangerous to children and the elderly, and it may be incredibly problematic for women who are pregnant. It can lead to central nervous system damage and even failure, as well as birth defects in unborn children. Mercury in water can be reduced if industries are held to higher standards of environmentally-safe practices and better work to ensure they prevent runoff:

- Animals: Mercury in water can poison fish, which in turn may poison other animals that eat them, as well as humans. Farm animals and even pets that drink water contaminated with mercury may suffer the effects of mercury poisoning, including damage to the central nervous system.

- Land and Plants: Plants exposed to high levels of mercury over long periods of time may suffer and die from an inability to properly process their nutrients.

- Water: The water itself must be filtered very carefully in order to remove all mercury build-up. Unfortunately, even some city water sources show high mercury levels.

- Air/Atmosphere: Mercury isn't able to be detected in the air.

Toxic Runoff

Toxic runoff has a lot of different causes, and unfortunately, it has a lot of different effects as well. Both industrial waste and waste from dumping (illegally or in landfills) can lead to a toxic runoff in huge amounts. When toxic chemicals are present in an industrial site or landfill, it only takes one heavy rain to start washing those chemicals away from the site and into the groundwater nearby. If there are creeks, ponds, lakes or rivers nearby, they may also be affected, especially if the contamination continues over a long period of time. Effects of toxic runoff on humans range from stomach upset to parasites and bacteria, and this may even cause cancer in some places. Toxic runoff can be prevented if landfills and industries are both more careful about practicing environmental safety when it comes to protecting the groundwater nearby.

- Animals: When animals drink water contaminated by toxic runoff, they are subjected to many of the same illnesses and bacteria as humans. If these animals include those that are later eaten by humans, illness and parasites can spread this way as well.

- Land and Plants: Plants exposed to high levels of mercury over long periods of time may suffer and die from an inability to properly process their nutrients.

- Water: It may be completely impossible to remove all the contamination from water that is exposed to toxic runoff. In some situations, the water cannot be salvaged.

- Air/Atmosphere: It is uncommon for toxic runoff to cause air pollution, but in some instances, it may, especially where toxic chemicals have very harsh smells.

Oil Spills

It's no secret that oil spills are a frequent cause of water contamination and pollution. You've probably heard a lot about the damage that takes place in the ocean when massive oil freighters or drills fail and leave spills in the water, but did you know that smaller-scale oil spills happen much more frequently in freshwater sources around the country? Anytime oil and gas are stored or transported, the potential for oil spills increases significantly. One accident could leave a truck overturned and leaking into a wetland, and this is unfortunately not all that uncommon. Oil spills can be remedied by turning to cleaner and safer sources of fuel, and they can be cut back on by using safer modes of storage and transportation.

- Animals: Animals suffer greatly from oil spills. Fish and mammals living in an area where a spill is present may be choked out completely from lack of access to fresh water. Oil spills contribute to extinction in some species.

- Land and Plants: Plants and soil suffer much the same fate as animals when oil spills take place. It can take years for plant life to regrow after damage from an oil spill.

- Water: In some instances, oil may never be able to be removed from the water. The Deepwater Horizon spill eventually moved from the ocean into wetlands and was never cleaned up.

- Air/Atmosphere: Very large oil spills may cause air pollution that can further affect both animals and humans in the area.

Algae Growth

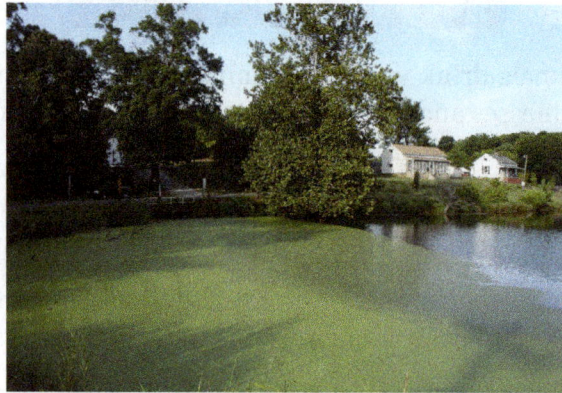

Algae is an important part of any water ecosystem. However, when algae growth gets out of control, problems very quickly arise. If you've ever had a fish tank at home, you've possibly seen the effects of algae overgrowth, and you might have added a few "algae eaters" to your tank to combat the problem. While these creatures also exist in nature, sometimes the algae growth gets too out of control for nature to be able to fix. This is often caused by nutrients in the water becoming too prevalent and upsetting the balance of the ecosystem. For example, when water is treated too heavily with fluoride, this nutrient overload can quickly lead to algae taking over. Treating water with other chemicals is a good way to reduce the risk of algae, but it can quickly come back to be a problem all over again if water treatment isn't very carefully maintained over time.

- Animals: When algae blooms grow too heavily in a body of water, the fish that live there quickly suffocate and die. Some animals may also die from drinking water that is heavily contaminated with algae.

- Land and Plants: Algae may eventually choke out plant life, especially in wetlands, and take over much like a weed.

- Water: It may be very difficult to ever remove an overabundance of algae from a water source, and it will require even more chemical treatment that may cause problems in the long run.

- Air/Atmosphere: Algae does not affect air pollution.

Acid Rain

Acid rain refers to a couple of different situations that both result in the same outcome. One of these situations involves very highly polluted water. When water is seriously polluted, it slowly evaporates into the atmosphere and leads to chemicals being picked up in the rain and dropped again elsewhere. However, the same thing can happen when the air is heavily polluted, especially in large cities or near sources of nuclear power. In these places, chemicals are drawn up into the atmosphere and carried, sometimes for miles, before being released once again as rain. Whichever way it happens, acid rain can cause damage to humans, property, and the environment.

- Animals: Acid rain may physically harm animals in very severe instances, and it may also lead to the destruction of animal habitats. It may also cause damage to human property, including stripping paint from vehicles and weakening roofs on homes.

- Land and Plants: Plants are seriously damaged by acid rain. They will die quickly in areas where this event occurs.

- Water: Water polluted by toxins continues to create more acid rain, which in turn creates more pollution.

- Air/Atmosphere: The air and atmosphere are seriously polluted when acid rain takes place.

Infectious Disease

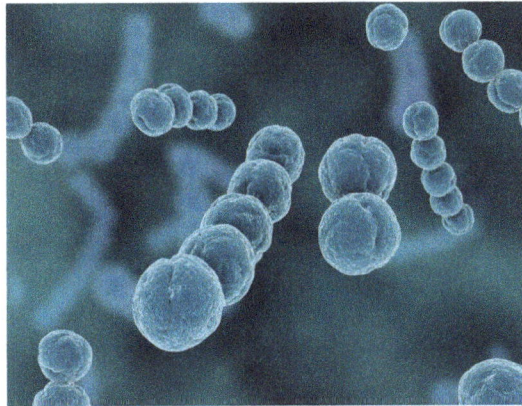

One of the biggest ways in which water pollution damages the environment is by contributing to the spread of infectious disease. This is one of the many outcomes of sewage water pollution and its environmental effects. Contaminated water quickly becomes a breeding ground for bacteria, parasites, and more that lead to disease almost immediately. This type of contamination can come from just about any source of water pollution, and it is one of the most common ways in which pollution affects humans in the United States as well as around the world. Up to seven million people in the United States alone get sick from drinking polluted water every year, and these illnesses can range from digestive disorders to cancer and more.

- Animals: Some animals can also become infected with certain types of disease and illness that may also affect humans. They are also very prone to picking up parasites from contaminated water, which they can then spread to humans.

- Land and Plants: Plants and soil aren't often directly affected by these illnesses, but they may still become contaminated, especially with such diseases as e. Coli, listeria, and salmonella.

- Water: Infected water must be very well treated before it is safe to drink. These diseases are some of the many effects of sewage water pollution, which must be treated even further when it occurs.

- Air/Atmosphere: Infectious disease doesn't cause atmospheric contamination, but many may be spread by air.

Agricultural Pesticides

Pesticides are used frequently in agricultural practices, as well as in residential communities. These pesticides are used with the good intention of keeping crops safe from harmful insects and other creatures that might cause them to fail. However, they are also very strong chemicals that quickly wash into groundwater and surface water, leading to serious pollution effects. When humans drink water polluted with pesticides, it can lead to cancer as well as developmental issues in children. It may also cause kidney and liver damage over time. This can be avoided by turning to environmentally-friendly ways to fight pests, such as using natural predators to keep them at bay.

- Animals: Animals may be directly affected by drinking water or eating food contaminated with pesticides. They may be poisoned and die easily when this happens.

- Land and Plants: The soil eventually suffers from overuse of pesticides, and may become infertile in some locations.

- Water: It can be very difficult to filter pesticides from water sources. These water sources continue to spread illness over long periods of time.

- Air/Atmosphere: Some pesticide chemicals may be spread through the air, causing respiratory issues in humans and animals.

Trash in Water

While it's true that most types of garbage dumped in and around water eventually lead to toxic runoff, there are other, more direct ways that trash in water sources can lead to damage to the environment. When large pieces of debris are present, especially in the ocean, the safety and health of the animals that live there are endangered. Many ocean creatures may swallow pieces of garbage and die from internal damage. If they don't die, they may carry the garbage in their bodies for a long time, and if they happen to be fish caught for human consumption, this could affect humans as well.

- Animals: Animals suffer the most from this type of pollution. Not just in the ocean, freshwater creatures can also be injured or killed by being trapped by garbage or eating trash dumped in the water.

- Land and Plants: Large pieces of debris can cause damage to plant life, especially in freshwater sources.

- Water: The water may remain dirty for years without human intervention to clean it up when trash is dumped there.

- Air/Atmosphere: Trash in the water doesn't lead to air pollution.

Thermal Pollution

Did you know that temperature can even be a type of pollution? In places where factories are present, industrial operations often lead to an increase in water temperatures in surrounding sources of freshwater. This may be imperceptible to humans, especially if it takes place slowly over a long period of time, but it can cause serious problems for animals and plants.

- Animals: Increased heat can cause fish to overpopulate, thinking it's time to spawn over again. However, it can also kill fish that are used to cooler temperatures. Oxygen in the water may be seriously limited by high temperatures.

- Land and Plants: Plants may eventually be choked out from a lack of oxygen in the water.

- Water: Water temperatures may never be able to be restored to normal again, depending on the source of the thermal pollution.

- Air/Atmosphere: Factories can sometimes also lead to atmospheric pollution that will, in turn, raise the temperature in the atmosphere as well.

Mining

As in the case of industries, mining can easily lead to toxic runoff that pollutes groundwater and surface water. However, the results are slightly different. When mining pollutes water, it usually causes the acidity in the water to rise, which can lead to a complete upset in an ecosystem. If humans drink this water, they may become very sick, depending on how acidic the water has grown. Safer and more environmentally-conscious mining practices must be instated in order to avoid this type of pollution.

- Animals: Animals die quickly when they drink water that is much too acidic. Fish are choked out in large numbers by these effects on water.

- Land and Plants: Plants cannot grow in areas where water is too acidic for them to absorb. While they may try for a while, eventually they will grow too weak to continue thriving in the environment.

- Water: Water may be very difficult to return to its previous level of acidity, especially if mining is still taking place.

- Air/Atmosphere: Mining may lead to some air pollution as well.

Rdioactivity

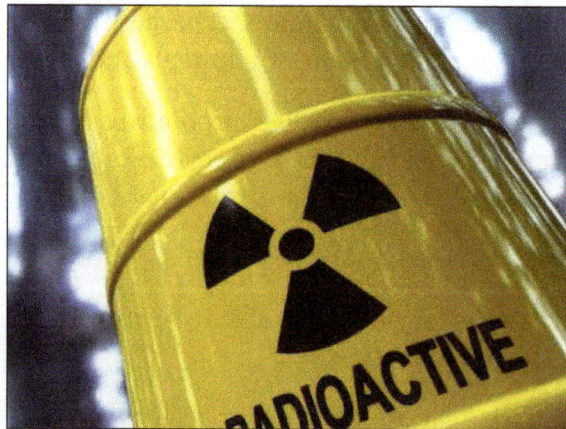

Areas where nuclear power plants are present may experience radioactivity pollution in their water sources. It may sound terrifying to imagine drinking water that is radioactive, but unfortunately, many people do this every day without realizing it. Drinking radioactive water can lead to many types of cancer, and it can lead to serious birth defects in unborn children as well. These genetic issues may be passed on to future generations, especially if the area remains contaminated by waste from these plants.

- Animals: Animals in areas where radioactive waste is present suffer genetic mutations and die much more quickly than they should. Fish that swim in water contaminated by radon are contaminated and die quickly. They should never be consumed.

- Land and Plants: The soil quickly becomes infertile in radioactively polluted areas, and it cannot sustain plant life. Plants may continue to grow but will be damaged and weak.

- Water: Radioactive water may be impossible to clean. This water is very unsafe to come into contact with.

- Air/Atmosphere: These chemicals can and are often absorbed into the atmosphere and lead to acid rain and air pollution.

Inorganic Chemicals

Inorganic chemicals are any chemicals that don't naturally occur in water sources. They may be put there by dumping, factory processes, agricultural use, or any number of other causes not previously listed. Depending on the type of pollutant present, there may be dire consequences for humans, including nerve damage and poisoning of the nervous system. These pollutants may include ammonia, cyanide, and hydrogen sulfide, among others.

- Animals: Fish gills may be burned and damaged by the presence of these chemicals in water. Animals may become sick and poisoned in much the same way as humans.

- Land and Plants: Plants may be burned by ammonia overload and may be stunted by any of these chemicals present in their water supplies.

- Water: These chemicals can be filtered from water in some instances.

- Air/Atmosphere: Sometimes, these chemicals can evaporate and lead to air pollution.

Organic Minerals

Organic minerals are present in nature in small amounts but may contaminate water supplies when human intervention exposes them more frequently to sources of water. Depending on the type of mineral present in water, it may lead to poisoning in humans, as well as to respiratory and circulatory dysfunction, birth defects, cancer, and nervous system damage. These pollutants may include arsenic, lead, iron, and more.

- Animals: Animals may suffer many of the same issues as humans, depending on the type of pollution present.

- Land and Plants: Plants may die or otherwise become sickly from exposure to these minerals. Farm crops cannot thrive in areas where this pollution is present.

- Water: Some of these naturally occurring minerals may be able to be filtered from water at low to medium levels.

- Air/Atmosphere: These pollutants do not lead to air or atmospheric pollution.

5

Diseases Caused by Water Pollution

Water pollution can adversely affect the health of humans by causing numerous diseases. Some of the water-borne diseases are cholera, dysentery, diarrhea, typhoid, jaundice, amoebiasis, malaria, etc. This chapter closely examines all these diseases caused by water pollution to provide an extensive understanding of the subject.

Water pollution involves the pollution of surface waters and/or groundwater, which may cause a series of diseases referred to as water pollution diseases. These could have serious health impacts. While we can control (to some extent) the water we drink, the pollution of our water streams may have long-term effects by reducing the "drinkable" water reserves of our planet. Additionally, the common filtration methods for water are not efficient for some of the new emerging contaminants – which are often not even tested for contaminants. Water pollution travels slower than air pollution but still may affect large areas.

Water may commonly be polluted by two main categories of pollutants (dissolved or suspended in water:

- Chemicals – including natural or man-made (xenobiotic) chemicals that gets into a water body (by being dissolved or dispersed in the water) and reaching concentrations that raise serious health concerns; note that, similar to the case of air pollutants, the presence of such pollutants in water is not always obvious and may not be detected by our senses. Common problematic chemicals getting into water are pesticides, chlorinated solvents, petroleum chemicals, mercury, PCBs, dioxins and other persisting organic pollutants; as well as any of the other tens of thousands of chemicals used in industrial processes.

- Living organisms (as long as they are induced by human activity; please note that some waters unaffected by human activity may still be naturally polluted with some of these organisms – in which case, the caused diseases may not be seen as pollution diseases:

    ◦ Pathogens – including a variety of living organisms (usually from animal waste) such as various species of viruses, bacteria, fungi and intestinal worms. Their presence in water, many times, remains unnoticed.

- Algae – some types of algae are toxic and may overgrow due to the presence of nitrates and phosphates in runoff water (especially agricultural runoff); such overgrowth is usually referred to as "red tides" or "brown tides". Their toxins may affect the food chain, including fish and birds, and ultimately humans. Oxygen depletion in polluted water is another serious problem responsible for killing fish all over the world.

While the most common water pollution diseases involve poisoning episodes affecting the digestive system and/or causing human infectious diseases, water pollution may cause a large variety of health diseases including:

- Infectious diseases caused by pathogens (usually microorganisms) from animal fecal origins, of which the most common occur in developing countries, including:

  - Typhoid.

  - Giardiasis.

  - Amoebiasis.

  - Ascariasis.

  - Hookworm.

- Diseases caused by polluted beach water, including:

  - Gastroenteritis.

  - Diarrhea.

  - Encephalitis.

  - Stomach cramps and aches.

  - Vomiting.

  - Hepatitis.

  - Respiratory infections.

- Liver damage and even cancer (due to DNA damage) – Caused by a series of chemicals (e.g., chlorinated solvents, MTBE).

- Kidney damage caused by a series of chemicals.

- Neurological problems - damage to the nervous system – Usually due to the presence of chemicals such as pesticides (e.g. DDT).

- Reproductive and endocrine damage including interrupted sexual development, inability to breed, degraded immune function, decreased fertility and increase in some types of cancers – caused by a series of chemicals including endocrine disruptors.

- Thyroid system disorders (a common cause is exposure to perchlorate, which is a chemical contaminating large water bodies such as the Colorado River).

- Increased water pollution creates breeding grounds for malaria-carrying mosquitoes, which kill 1.2-2.7 million people a year.

- A series of less serious health effects could be associated to bathing in contaminated water (i.e. polluted beach water) including:

    ◦ Rashes.

    ◦ Earaches.

    ◦ Pink eyes.

Water pollution can affect us:

- Directly – through consumption or bathing in a polluted stream (such as consumption of municipal water, as well as bathing in polluted lakes or beach water).

- Indirectly – through the consumption of vegetables irrigated with contaminated water, as well as of fish or other animals that live in the polluted water or consume animals grown in the polluted water. This is many times more dangerous than being directly affected through consumption of water, because some pollutants bioaccumulate in fish and living organisms (their concentration in fish could be several orders of magnitude higher than their water concentration). Additionally, the toxins from the brown tide are strong and can travel via air, affecting homeowners close to the beach.

The most common ways of polluting the water include:

- Waste disposal:

    ◦ Directly into water streams.

    ◦ Onto the soil from which contaminants may leak into the groundwater below.

- Urban and agricultural runoff.

- Animal waste could also add dangerous pathogens (usually microbial groups, viruses and intestinal parasites) into the water.

- From air via acid rain - water can get polluted with air contaminants (that have sometimes traveled long distances – such as the case of Hg) that reach the land and water via acid rain. During precipitation, air pollutants may get dissolved in the water drops and, as a result, they may acidify the water - which is why polluted rainwater is referred to as "acid rain".

According to a Cornell University study, water pollution accounts for 80% of all infectious diseases. According to the same study, unsanitary living conditions account for more than 5 million deaths a year.

Cholera

Cholera is a bacterial disease usually spread through contaminated water. Cholera causes severe diarrhea and dehydration. Left untreated, cholera can be fatal in a matter of hours, even in previously healthy people.

Modern sewage and water treatment have virtually eliminated cholera in industrialized countries. The last major outbreak in the United States occurred in 1911. But cholera is still present in Africa, Southeast Asia and Haiti. The risk of cholera epidemic is highest when poverty, war or natural disasters force people to live in crowded conditions without adequate sanitation.

Cholera is easily treated. Death results from severe dehydration that can be prevented with a simple and inexpensive rehydration solution.

Symptoms

Most people exposed to the cholera bacterium (Vibrio cholerae) don't become ill and never know they've been infected. Yet because they shed cholera bacteria in their stool for seven to 14 days, they can still infect others through contaminated water. Most symptomatic cases of cholera cause mild or moderate diarrhea that's often hard to distinguish from diarrhea caused by other problems.

Only about 1 in 10 infected people develops more-serious signs and symptoms of cholera, usually within a few days of infection.

Symptoms of Cholera Infection may include:

- Diarrhea: Cholera-related diarrhea comes on suddenly and may quickly cause dangerous fluid loss — as much as a quart (about 1 liter) an hour. Diarrhea due to cholera often has a pale, milky appearance that resembles water in which rice has been rinsed (rice-water stool).

- Nausea and vomiting: Occurring especially in the early stages of cholera, vomiting may persist for hours at a time.

- Dehydration: Dehydration can develop within hours after the onset of cholera symptoms. Depending on how many body fluids have been lost, dehydration can range from mild to severe. A loss of 10 percent or more of total body weight indicates severe dehydration.

Signs and symptoms of cholera dehydration include irritability, lethargy, sunken eyes, a dry mouth, extreme thirst, dry and shriveled skin that's slow to bounce back when pinched into a fold, little or no urine output, low blood pressure, and an irregular heartbeat (arrhythmia).

Dehydration may lead to a rapid loss of minerals in your blood (electrolytes) that maintain the balance of fluids in your body. This is called an electrolyte imbalance.

Electrolyte Imbalance

An electrolyte imbalance can lead to serious signs and symptoms such as:

- Muscle cramps: These result from the rapid loss of salts such as sodium, chloride and potassium.

- Shock: This is one of the most serious complications of dehydration. It occurs when low blood volume causes a drop in blood pressure and a drop in the amount of oxygen in your body. If untreated, severe hypovolemic shock can cause death in a matter of minutes.

Signs and Symptoms of Cholera in Children

In general, children with cholera have the same signs and symptoms adults do, but they are particularly susceptible to low blood sugar (hypoglycemia) due to fluid loss, which may cause:

- An altered state of consciousness.

- Seizures.

- Coma.

The risk of cholera is slight in industrialized nations, and even in endemic areas you're not likely to become infected if you follow food safety recommendations. Still, sporadic cases of cholera occur throughout the world. If you develop severe diarrhea after visiting an area with active cholera, see your doctor.

If you have diarrhea, especially severe diarrhea, and think you may have been exposed to cholera, seek treatment right away. Severe dehydration is a medical emergency that requires immediate care regardless of the cause.

Causes

A bacterium called Vibrio cholerae causes cholera infection. However, the deadly effects of the disease are the result of a potent toxin called CTX that the bacterium produce in the small intestine. CTX binds to the intestinal walls, where it interferes with the normal flow of sodium and chloride. This causes the body to secrete enormous amounts of water, leading to diarrhea and a rapid loss of fluids and salts (electrolytes).

Contaminated water supplies are the main source of cholera infection, although raw shellfish, uncooked fruits and vegetables, and other foods also can harbor V. cholerae.

Cholera bacteria have two distinct life cycles — one in the environment and one in humans.

Cholera Bacteria in the Environment

Cholera bacteria occur naturally in coastal waters, where they attach to tiny crustaceans called copepods. The cholera bacteria travel with their hosts, spreading worldwide as the crustaceans follow their food source — certain types of algae and plankton that grow explosively when water temperatures rise. Algae growth is further fueled by the urea found in sewage and in agricultural runoff.

Cholera Bacteria in People

When humans ingest cholera bacteria, they may not become sick themselves, but they still pass the bacteria in their stool. When human feces contaminate food and water supplies, both can serve as ideal breeding grounds for the cholera bacteria.

Because more than a million cholera bacteria — approximately the amount you'd find in a glass of contaminated water — are needed to cause illness, cholera usually isn't transmitted through casual person-to-person contact.

The most common sources of cholera infection are standing water and certain types of food, including seafood, raw fruits and vegetables, and grains:

- Surface or well water: Cholera bacteria can lie dormant in water for long periods, and contaminated public wells are frequent sources of large-scale cholera outbreaks. People living in crowded conditions without adequate sanitation are especially at risk of cholera.

- Seafood: Eating raw or undercooked seafood, especially shellfish, that originates from certain locations can expose you to cholera bacteria. Most recent cases of cholera occurring in the United States have been traced to seafood from the Gulf of Mexico.

- Raw fruits and vegetables: Raw, unpeeled fruits and vegetables are a frequent source of cholera infection in areas where cholera is endemic. In developing nations, uncomposted manure fertilizers or irrigation water containing raw sewage can contaminate produce in the field.

- Grains: In regions where cholera is widespread, grains such as rice and millet that are contaminated after cooking and allowed to remain at room temperature for several hours become a medium for the growth of cholera bacteria.

Risk Factors

Everyone is susceptible to cholera, with the exception of infants who derive immunity from nursing mothers who have previously had cholera. Still, certain factors can make you more vulnerable to the disease or more likely to experience severe signs and symptoms. Risk factors for cholera include:

- Poor sanitary conditions: Cholera is more likely to flourish in situations where a sanitary environment — including a safe water supply — is difficult to maintain. Such conditions are common to refugee camps, impoverished countries, and areas devastated by famine, war or natural disasters.

- Reduced or nonexistent stomach acid (hypochlorhydria or achlorhydria): Cholera bacteria can't survive in an acidic environment, and ordinary stomach acid often serves as a first line defense against infection. But people with low levels of stomach acid — such as children, older adults, and people who take antacids, H-2 blockers or proton pump inhibitors — lack this protection, so they're at greater risk of cholera.

- Household exposure: You're at significantly increased risk of cholera if you live with someone who has the disease.

- Type O blood: For reasons that aren't entirely clear, people with type O blood are twice as likely to develop cholera compared with people with other blood types.

- Raw or undercooked shellfish: Although large-scale cholera outbreaks no longer occur in industrialized nations, eating shellfish from waters known to harbor the bacteria greatly increases your risk.

Complications

Cholera can quickly become fatal. In the most severe cases, the rapid loss of large amounts of fluids and electrolytes can lead to death within two to three hours. In less extreme situations, people who don't receive treatment may die of dehydration and shock hours to days after cholera symptoms first appear.

Although shock and severe dehydration are the most devastating complications of cholera, other problems can occur, such as:

- Low blood sugar (hypoglycemia): Dangerously low levels of blood sugar (glucose) — the body's main energy source — may occur when people become too ill to eat. Children are at greatest risk of this complication, which can cause seizures, unconsciousness and even death.

- Low potassium levels (hypokalemia): People with cholera lose large quantities of minerals, including potassium, in their stools. Very low potassium levels interfere with heart and nerve function and are life-threatening.

- Kidney (renal) failure: When the kidneys lose their filtering ability, excess amounts of fluids, some electrolytes and wastes build up in your body — a potentially life-threatening condition. In people with cholera, kidney failure often accompanies shock.

Prevention

Cholera is rare in the United States with the few cases related to travel outside the U.S. or to contaminated and improperly cooked seafood from the Gulf Coast waters.

If you're traveling to cholera-endemic areas, your risk of contracting the disease is extremely low if you follow these precautions:

- Wash hands with soap and water frequently, especially after using the toilet and before handling food. Rub soapy, wet hands together for at least 15 seconds before rinsing. If soap and water aren't available, use an alcohol-based hand sanitizer.

- Drink only safe water, including bottled water or water you've boiled or disinfected yourself. Use bottled water even to brush your teeth. Hot beverages are generally safe, as are canned or bottled drinks, but wipe the outside before you open them. Avoid adding ice to your beverages unless you made it yourself using safe water.

- Eat food that's completely cooked and hot and avoid street vendor food, if possible. If you do buy a meal from a street vendor, make sure it's cooked in your presence and served hot

- Avoid sushi, as well as raw or improperly cooked fish and seafood of any kind.

- Stick to fruits and vegetables that you can peel yourself, such as bananas, oranges and avocados. Stay away from salads and fruits that can't be peeled, such as grapes and berries.

- Be wary of dairy foods, including ice cream, which is often contaminated, and unpasteurized milk.

Cholera Vaccine

For adults traveling to areas affected by cholera, a vaccine is now available in the United States. The Food and Drug Administration recently approved Vaxchora, a vaccine for the prevention of cholera. It is a liquid dose taken by mouth at least 10 days before travel.

A few countries offer oral vaccines as well. Contact your doctor or local office of public health for more information about these vaccines. Keep in mind that no country requires immunization against cholera as a condition for entry.

Dysentery

Dysentery is an inflammatory disease of the intestine, especially of the colon, which always results in severe diarrhea and abdominal pains. Other symptoms may include fever and a feeling of incomplete defecation. The disease is caused by several types of infectious pathogens such as bacteria, viruses and parasites.

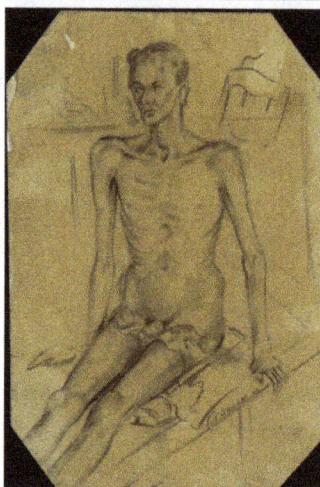

A person with dysentery in a Burma hospital, 1943	
Specialty	Infectious disease
Symptoms	Bloody diarrhea, abdominal pain, fever
Causes	Usually Shigella or Entamoeba histolytica

Signs and Symptoms

The most common form of dysentery is bacillary dysentery, which is typically a mild sickness, causing symptoms normally consisting of mild gut pains and frequent passage of stool or diarrhea. Symptoms normally present themselves after 1-3 days, and are usually no longer present after a week. The frequency of urges to defecate, the large volume of liquid feces ejected, and the presence of blood, mucus or pus depends on the pathogen causing the disease. Temporary lactose intolerance can occur, as well. In some caustic occasions severe abdominal cramps, fever, shock and delirium can all be symptoms.

In extreme cases, dysentery patients may pass more than one liter of fluid per hour. More often, individuals will complain of intense abdominal pains and severe diarrhea with blood or mucus, accompanied by rectal pain and low-grade fever. Rapid weight loss and generalized muscle aches sometimes also accompany dysentery, while nausea and vomiting are rare. On rare occasions, the amoebic parasite will invade the body through the bloodstream and spread beyond the intestines. In such cases, it may more seriously infect other organs such as the brain, lungs, and most commonly the liver.

Mechanism

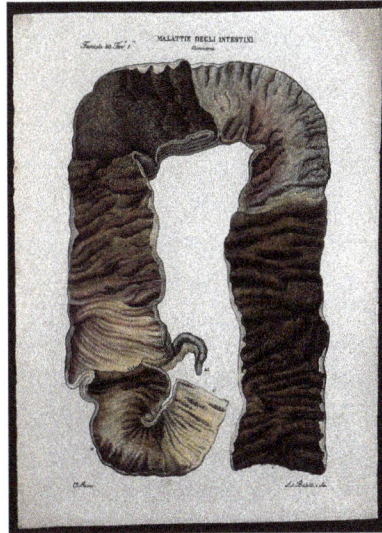

Cross-section of diseased intestines. Colored lithograph c. 1843.

Dysentery results from viral, bacterial, or parasitic infections. These pathogens typically reach the large intestine after entering orally, through ingestion of contaminated food or water, oral contact with contaminated objects or hands, and so on.

Each specific pathogen has its own mechanism or pathogenesis, but in general, the result is damage to the intestinal linings, leading to the inflammatory immune responses. This can cause elevated physical temperature, painful spasms of the intestinal muscles (cramping), swelling due to fluid leaking from capillaries of the intestine (edema) and further tissue damage by the body's immune cells and the chemicals, called cytokines, which are released to fight the infection. The result can be impaired nutrient absorption, excessive water and mineral loss through the stools due to breakdown of the control mechanisms in the intestinal tissue that normally remove water from the stools, and in severe cases, the entry of pathogenic organisms into the bloodstream. Anemia may also arise due to the blood loss through diarrhea.

Bacterial infections that cause bloody diarrhea are typically classified as being either invasive or toxogenic. Invasive species cause damage directly by invading into the mucosa. The toxogenic species do not invade, but cause cellular damage by secreting toxins, resulting in bloody diarrhea. This is also in contrast to toxins that cause watery diarrhea, which usually do not cause cellular damage, but rather they take over cellular machinery for a portion of life of the cell.

Some microorganisms – for example, bacteria of the genus *Shigella* – secrete substances known as cytotoxins, which kill and damage intestinal tissue on contact. Shigella is thought to cause bleeding due to invasion rather than toxin, because even non-toxogenic strains can cause dysentery, but E. coli with shiga-like toxins do not invade the intestinal mucosa, and are therefore toxin dependent. Viruses directly attack the intestinal cells, taking over their metabolic machinery to make copies of themselves, which leads to cell death.

Definitions of dysentery can vary by region and by medical specialty. The U. S. Centers for Disease Control and Prevention (CDC) limits its definition to "diarrhea with visible blood". Others define the term more broadly. These differences in definition must be taken into account when defining mechanisms. For example, using the CDC definition requires that intestinal tissue be so severely damaged that blood vessels have ruptured, allowing visible quantities of blood to be lost with defecation. Other definitions require less specific damage.

Amoebic Dysentery

Amoebiasis, also known as amoebic dysentery, is caused by an infection from the amoeba *Entamoeba histolytica*, which is found mainly in tropical areas. Proper treatment of the underlying infection of amoebic dysentery is important; insufficiently treated amoebiasis can lie dormant for years and subsequently lead to severe, potentially fatal, complications.

When amoebae inside the bowel of an infected person are ready to leave the body, they group together and form a shell that surrounds and protects them. This group of amoebae is known as a cyst, which is then passed out of the person's body in the feces and can survive outside the body. If hygiene standards are poor — for example, if the person does not dispose of the feces hygienically — then it can contaminate the surroundings, such as nearby food and water. If another person then eats or drinks food or water that has been contaminated with feces containing the cyst, that person will also become infected with the amoebae. Amoebic dysentery is particularly common in parts of the world where human feces are used as fertilizer. After entering the person's body through the mouth, the cyst travels down into the stomach. The amoebae inside the cyst are protected from the stomach's digestive acid. From the stomach, the cyst travels to the intestines, where it breaks open and releases the amoebae, causing the infection. The amoebae can burrow into the walls of the intestines and cause small abscesses and ulcers to form. The cycle then begins again.

Bacillary Dysentery

Dysentery may also be caused by shigellosis, an infection by bacteria of the genus *Shigella*, and is then known as bacillary dysentery (or Marlow syndrome). The term *bacillary dysentery* etymologically might seem to refer to any dysentery caused by any bacilliform bacteria, but its meaning is restricted by convention to *Shigella* dysentery.

Other Bacterial Diarrhea

Some strains of *Escherichia coli* cause bloody diarrhea. The typical culprits are enterohemorrhagic *Escherichia coli*, of which O157:H7 is the best known.

Diagnosis

A clinical diagnosis may be made by taking a history and doing a brief examination. Treatment is usually started without or before confirmation by laboratory analysis.

Physical Exam

The mouth, skin, and lips may appear dry due to dehydration. Lower abdominal tenderness may also be present.

Stool and Blood tests

Cultures of stool samples are examined to identify the organism causing dysentery. Usually, several samples must be obtained due to the number of amoebae, which changes daily. Blood tests can be used to measure abnormalities in the levels of essential minerals and salts.

Treatment

Dysentery is managed by maintaining fluids by using oral rehydration therapy. If this treatment cannot be adequately maintained due to vomiting or the profuseness of diarrhea, hospital admission may be required for intravenous fluid replacement. In ideal situations, no antimicrobial therapy should be administered until microbiological microscopy and culture studies have established the specific infection involved. When laboratory services are not available, it may be necessary to administer a combination of drugs, including an amoebicidal drug to kill the parasite, and an antibiotic to treat any associated bacterial infection.

If shigellosis is suspected and it is not too severe, letting it run its course may be reasonable — usually less than a week. If the case is severe, antibiotics such as ciprofloxacin or TMP-SMX may be useful. However, many strains of *Shigella* are becoming resistant to common antibiotics, and effective medications are often in short supply in developing countries. If necessary, a doctor may have to reserve antibiotics for those at highest risk for death, including young children, people over 50, and anyone suffering from dehydration or malnutrition.

Amoebic dysentery is often treated with two antimicrobial drugs such as metronidazole and paromomycin or iodoquinol.

The seed, leaves, and bark of the kapok tree have been used in traditional medicine by indigenous peoples of the rainforest regions in the Americas, west-central Africa, and Southeast Asia to treat this disease. *Bacillus subtilis* was marketed throughout America and Europe from 1946 as an immunostimulatory aid in the treatment of gut and urinary tract diseases such as rotavirus and *Shigella*, but declined in popularity after the introduction of consumer antibiotics.

Prognosis

With correct treatment, most cases of amoebic and bacterial dysentery subside within 10 days, and most individuals achieve a full recovery within two to four weeks after beginning proper treatment. If the disease is left untreated, the prognosis varies with the immune status of the individual patient and the severity of disease. Extreme dehydration can delay recovery and significantly raises the risk for serious complications.

Diarrhea

Diarrhea is one of the most common health complaints. It can range from a mild, temporary condition, to a potentially life-threatening one.

Globally, an estimated 2 billion cases of diarrheal disease occur each year, and 1.9 million children under the age of 5 years, mostly in developing countries, die from diarrhea.

Diarrhea is characterized by abnormally loose or watery stools.

Some people frequently pass stools, but they are of normal consistency. This is not diarrhea. Similarly, breastfed babies often pass loose, pasty stools. This is normal. It is not diarrhea.

Fast Facts on Diarrhea

Here are some key points about diarrhea. More detail and supporting information is in the body of this article:

- Most cases of diarrhea are caused by bacteria, viruses, or parasites.

- Inflammatory bowel diseases (IBD) including Crohn's disease and ulcerative colitis can cause chronic diarrhea.

- Antidiarrheal medications can reduce diarrheal output and zinc supplement is effective in children.

- Some nutritional and probiotic interventions may help prevent diarrhea.

Causes

Most cases of diarrhea are caused by an infection in the gastrointestinal tract. The microbes responsible for this infection include:

- Bacteria.

- Viruses.

- Parasitic organisms.

The most commonly identified causes of acute diarrhea in the United States are the bacteria Salmonella, Campylobacter, Shigella, and Shiga-toxin-producing Escherichia coli.

Correcting Dehydration is the Priority of Diarrhea Treatment.

Some cases of chronic diarrhea are called "functional" because a clear cause cannot be found. In the developed world, irritable bowel syndrome (IBS) is the most common cause of functional diarrhea IBS is a complex of symptoms. There is cramping abdominal pain and altered bowel habits, including diarrhea, constipation, or both.

Inflammatory bowel disease (IBD) is another cause of chronic diarrhea. It is a term used to describe either ulcerative colitis or Crohn's disease. There is often blood in the stool in both conditions.

Other major causes of chronic diarrhea include:

- Microscopic colitis: This is a persistent diarrhea that usually affects older adults, often during the night.

- Malabsorptive and maldigestive diarrhea: The first is caused by impaired nutrient absorption, the second by impaired digestive function. Celiac disease is one example.

- Chronic infections: A history of travel or antibiotic use can be clues to chronic diarrhea. Various bacteria and parasites can be the cause.

- Drug-induced diarrhea: Laxatives and other drugs, including antibiotics, can trigger diarrhea.

- Endocrine causes: Sometimes hormonal factors cause diarrhea, for example, in the case of Addison disease and carcinoid tumors.

- Cancer causes: Neoplastic diarrhea is associated with a number of gut cancers.

Symptoms

Symptoms of diarrhea can include bloating, thirst, and weight loss.

Diarrhea refers to watery stools, but it may be accompanied by other symptoms.

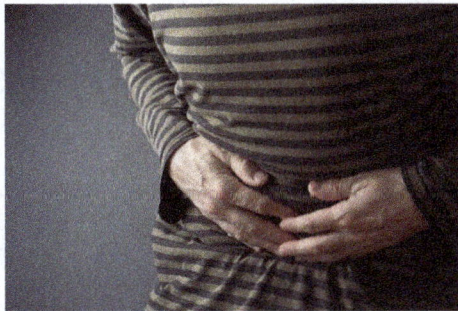

Symptoms of Diarrhea can Include Bloating, Thirst, and Weight Loss.

These include:

- Stomach pain.
- Abdominal cramps.
- Bloating.
- Thirst.
- Weight loss.
- Fever.

Diarrhea is a symptom of other conditions, some of which can be serious.

Other possible symptoms are:

- Blood or pus in the stools.

- Persistent vomiting.

- Dehydration.

If these accompany diarrhea, or if the diarrhea is chronic, it may indicate a more serious illness.

Complications

Two potentially serious complications of diarrhea are:

- Dehydration, with acute or chronic diarrhea.

- Malabsorption, with chronic diarrhea.

Diarrhea can also be a sign of a wide range of underlying chronic conditions. These conditions need to be diagnosed treated to prevent further problems.

Tests and Diagnosis

The doctor will ask about the symptoms and about any current medications, past medical history, and other medical conditions.

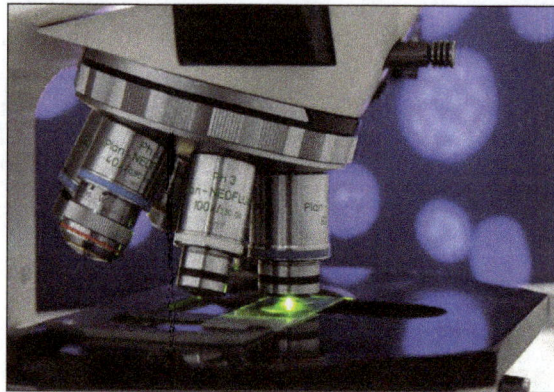

Parasites or their Eggs can be seen under a Microscope.

They will also ask:

- When the problem started.

- How frequent the stools are.

- Whether blood is present in the stool.

- Whether there has been vomiting.

- Whether the stools are watery, mucus- or pus-filled, and how much stool there is.

The doctor will also look for signs of dehydration Severe dehydration can be fatal if treatment with rehydration therapy is not given urgently.

Tests for Diarrhea

Most cases of diarrhea resolve without treatment, and a doctor will often be able to diagnose the problem without tests.

However, in more severe cases, a stool test may be needed, especially if the patient is very young or old Further tests may also be recommended if the patient:

- Has signs of fever or dehydration.

- Has stools with blood or pus.

- Has severe pain.

- Has low blood pressure.

- Has a weakened immune system.

- Has recently traveled to places outside Western Europe, North America, Australia, and New Zealand.

- Has recently received antibiotics or been in hospital.

- Has diarrhea persisting for more than 1 week.

if a person has chronic or persistent diarrhea, the doctor will order tests according to the suspected underlying cause.

These may include the following investigations:

- Full blood count: Anemia or a raised platelet count will suggest inflammation.

- Liver function tests: This will include testing albumin levels.

- Tests for malabsorption: These will check the absorption of calcium, vitamin B-12, and folate. They will also assess iron status and thyroid function.

- Erythrocyte sedimentation rate (ESR) and C-reactive protein (CRP) Raised levels may indicate inflammatory bowel disease (IBD).

- Testing for antibodies: This may detect celiac disease.

When to see a Doctor

Diarrhea often resolves without specialist medical treatment, but sometimes it is important to seek a doctor's help.

Infants under 1 year should see a doctor if they have had 6 bouts of diarrhea or 3 bouts of vomiting within 24 hours.

Children over 1 year should see a doctor if they have had 6 episodes or more of diarrhea in 24 hours, or if there is diarrhea and vomiting at the same time.

It is important to seek medical help in the following cases:

- Persistent vomiting.

- Persistent diarrhea.

- Dehydration.

- Significant weight loss.

- Pus in the stool.

- Blood in the stool, which may turn the stool black.

Anyone who experiences diarrhea after surgery, after spending time in hospital, or after using antibiotics, should seek medical assistance.

Adults whose sleep is persistently disturbed by diarrhea may be able to get help to solve this problem.

Treatment

Mild cases of acute diarrhea may resolve without treatment. Persistent or chronic diarrhea will be diagnosed and any underlying causes will be treated in addition to the symptoms of diarrhea.

Dehydration

For all cases of diarrhea, rehydration is key:

- Fluids can be replaced by simply drinking more fluids, or they can be received intravenously in severe cases. Children and older people are more vulnerable to dehydration.

- Oral rehydration solution or salts (ORS) refers to water that contains salt and glucose. It is absorbed by the small intestine to replace the water and electrolytes lost in the stool. In developing countries, ORS costs just a few cents. The World Health Organization (WHO) says ORS can safely and effectively treat over 90 percent of non-severe diarrhea cases.

- Oral rehydration products, such as Oralyte and Rehydralyte, are available commercially. Zinc supplementation may reduce the severity and duration of diarrhea in children. Various products are available to purchase online.

Antidiarrheal Medication

Over-the-counter (OTC) antidiarrheal medicines are also available:

- Loperamide, or Imodium, is an antimotility drug that reduces stool passage. Loperamide and Imodium are both available to purchase over-the-counter or online.

- Bismuth subsalicylate, for example, Pepto-Bismol, reduces diarrheal stool output in adults and children. It can also be used to prevent traveler's diarrhea. The can be bought online as well as over-the-counter.

There is some concern that antidiarrheal medications could prolong bacterial infection by reducing the removal of pathogens through stools.

Antibiotics

Antibiotics are only used to treat diarrhea caused by a bacterial infection. If the cause is a certain medication, switching to another drug might be possible.

Diet

Nutritionists from Stanford Health Care offer some nutritional tips for diarrhea:

- Sip on clear, still liquids such as fruit juice without added sugar.

- After each loose stool, replace lost fluids with at least one cup of liquid.

- Do most of the drinking between, not during meals.

- Consume high-potassium foods and liquids, such as diluted fruit juices, potatoes without the skin, and bananas.

- Consume high-sodium foods and liquids, such as broths, soups, sports drinks, and salted crackers.

Other advice from the nutritionists is to:

- Eat foods high in soluble fiber, such as banana, oatmeal and rice, as these help thicken the stool.

- Limit foods that may make diarrhea worse, such as creamy, fried, and sugary foods.

Foods and drinks that might make the diarrhea worse include:

- Sugar-free gum, mints, sweet cherries, and prunes.

- Caffeinated drinks and medication.

- Fructose in high amounts, from fruit juices, grapes, honey, dates, nuts, figs, soft drinks, and prunes.

- Lactose in dairy products.

- Magnesium.

- Olestra, or Olean, a fat substitute.

Probiotics

There is mixed evidence for the role of probiotics in diarrhea. They may help prevent traveler's diarrhea. In children, there is evidence that they might reduce diarrheal illness by 1 day.

Antibiotic-associated diarrhea might be reduced by the use of probiotics, as may diarrhea related to Clostridium difficile, although the evidence is mixed.

People should ask their doctor for advice, as there are numerous strains. The strain most studied for antibiotic-associated diarrhea are probiotics based on Lactobacillus rhamnosus and Saccharomyces boulardii.

Probiotics to help with Clostridium difficile and antibiotic diarrheas were investigated in a trial published in The Lancet. They found no evidence that a multi-strain preparation of bacteria was effective in preventing these conditions, calling for a better understanding of the development of antibiotic-associated diarrhea.

Probiotics are available in capsules, tablets, powders, and liquids, and may be purchase online.

Prevention

In developing countries, prevention of diarrhea may be more challenging due to dirty water and poor sanitation.

The following can help prevent diarrhea:

- Clean and safe drinking water.

- Good sanitation systems, for example, waste water and sewage.

- Good hygiene practices, including handwashing with soap after defecation, after cleaning a child who has defecated, after disposing of a child's stool, before preparing food, and before eating.

- Breastfeeding for the first 6 months of life.

- Education on the spread of infection.

There is evidence that interventions from public health bodies to promote hand washing can cut diarrhea rates by about one-third.

Typhoid Fever

Typhoid fever, also known simply as typhoid, is a bacterial infection due to specific type of *Salmonella* that causes symptoms. Symptoms may vary from mild to severe, and usually begin 6 to 30 days after exposure. Often there is a gradual onset of a high fever over several days. This is commonly accompanied by weakness, abdominal pain, constipation, headaches, and mild vomiting. Some people develop a skin rash with rose colored spots. In severe cases, people may experience confusion. Without treatment, symptoms may last weeks or months. Diarrhea is uncommon. Other people may carry the bacterium without being affected; however, they are still able to spread the disease to others. Typhoid fever is a type of enteric fever, along with paratyphoid fever.

The cause is the bacterium *Salmonella enterica* subsp. *enterica* growing in the intestines and blood. Typhoid is spread by eating or drinking food or water contaminated with the feces of an infected person. Risk factors include poor sanitation and poor hygiene. Those who travel in the developing world are also at risk. Only humans can be infected. Symptoms are similar to those of many other infectious diseases. Diagnosis is by either culturing the bacteria or detecting their DNA in the blood, stool, or bone marrow. Culturing the bacterium can be difficult. Bone-marrow testing is the most accurate.

Typhoid fever	
Other names	Slow fever, typhoid
	Rose spots on the chest of a person with typhoid fever
Specialty	Infectious disease
Symptoms	Fever, abdominal pain, headache, rash
Usual onset	6–30 days after exposure
Causes	Salmonella enterica subsp. enterica (spread by food or water contaminated with feces)
Risk factors	Poor sanitation, poor hygiene.
Diagnostic method	Bacterial culture, DNA detection
Differential diagnosis	Other infectious diseases
Prevention	Typhoid vaccine, handwashing
Treatment	Antibiotics
Frequency	12.5 million (2015)
Deaths	149,000 (2015)

A typhoid vaccine can prevent about 40 to 90% of cases during the first two years. The vaccine may have some effect for up to seven years. For those at high risk or people traveling to areas where the disease is common, vaccination is recommended. Other efforts to prevent the disease include providing clean drinking water, good sanitation, and handwashing. Until an individual's infection is confirmed as cleared, the individual should not prepare food for others. The disease is treated with antibiotics such as azithromycin, fluoroquinolones, or third-generation cephalosporins. Resistance to these antibiotics has been developing, which has made treatment of the disease more difficult.

In 2015, 12.5 million new cases worldwide were reported. The disease is most common in India. Children are most commonly affected. Rates of disease decreased in the developed world in the 1940s as a result of improved sanitation and use of antibiotics to treat the disease. Each year in the United States, about 400 cases are reported and the disease occurs in an estimated 6,000 people. In 2015, it resulted in about 149,000 deaths worldwide – down from 181,000 in 1990 (about 0.3% of the global total). The risk of death may be as high as 20% without treatment. With treatment, it is between 1 and 4%. Typhus is a different disease. However, the name typhoid means "resembling typhus" due to the similarity in symptoms.

Signs and Symptoms

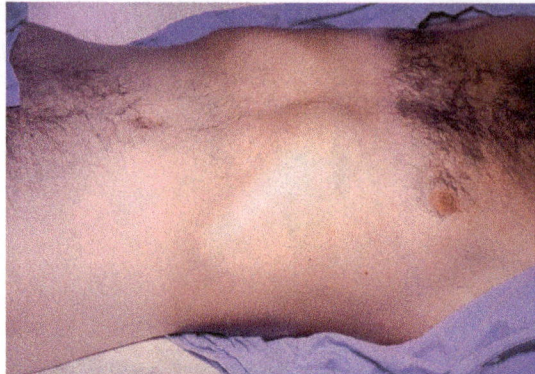

Rose spots on chest of a person with typhoid fever.

Classically, the progression of untreated typhoid fever is divided into four distinct stages, each lasting about a week. Over the course of these stages, the patient becomes exhausted and emaciated.

- In the first week, the body temperature rises slowly, and fever fluctuations are seen with relative bradycardia (Faget sign), malaise, headache, and cough. A bloody nose (epistaxis) is seen in a quarter of cases, and abdominal pain is also possible. A decrease in the number of circulating white blood cells (leukopenia) occurs with eosinopenia and relative lymphocytosis; blood cultures are positive for *Salmonella enterica* subsp. *enterica* or *S. paratyphi*. The Widal test is usually negative in the first week.

- In the second week, the person is often too tired to get up, with high fever in plateau around 40 °C (104 °F) and bradycardia (sphygmothermic dissociation or Faget sign), classically with a dicrotic pulse wave. Delirium can occur, where the patient is often calm, but sometimes becomes agitated. This delirium has lead to typhoid receiving the nickname "nervous fever". Rose spots appear on the lower chest and abdomen in around a third of patients. Rhonchi (rattling breathing sounds) are heard in the base of the lungs. The abdomen is distended and painful in the right lower quadrant, where a rumbling sound can be heard. Diarrhea can occur in this stage, but constipation is also common. The spleen and liver are enlarged (hepatosplenomegaly) and tender, and liver transaminases are elevated. The Widal test is strongly positive, with antiO and antiH antibodies. Blood cultures are sometimes still positive at this stage. The major symptom of this fever is that it usually rises in the afternoon up to the first and second week.

- In the third week of typhoid fever, a number of complications can occur:

  ○ Intestinal haemorrhage due to bleeding in congested Peyer's patches occurs; this can be very serious, but is usually not fatal.

  ○ Intestinal perforation in the distal ileum is a very serious complication and is frequently fatal. It may occur without alarming symptoms until septicaemia or diffuse peritonitis sets in.

- ◦ Encephalitis.

- ◦ Respiratory diseases such as pneumonia and acute bronchitis.

- ◦ Neuropsychiatric symptoms (described as "muttering delirium" or "coma vigil"), with picking at bedclothes or imaginary objects.

- ◦ Metastatic abscesses, cholecystitis, endocarditis, and osteitis.

- ◦ The fever is still very high and oscillates very little over 24 hours. Dehydration ensues, and the patient is delirious (typhoid state). One-third of affected individuals develop a macular rash on the trunk.

- ◦ Platelet count goes down slowly and the risk of bleeding rises.

- • By the end of third week, the fever starts subsiding.

Causes

A 1939 conceptual illustration showing various ways that typhoid bacteria can contaminate a water well (center).

Bacteria

The Gram-negative bacterium that causes typhoid fever is *Salmonella enterica* subsp. *enterica*. The two main types of the subspecies *enterica* are ST1 and ST2, based on MLST subtyping scheme, which are currently widespread globally.

Transmission

Unlike other strains of *Salmonella*, no animal carriers of typhoid are known. Humans are the only known carriers of the bacteria. *S. e.* subsp. *enterica* is spread through the fecal-oral route from individuals who are currently infected and from asymptomatic carriers of the bacteria. An asymptomatic human carrier is an individual who is still excreting typhoid bacteria in their stool a year after the acute stage of the infection.

Diagnosis

Diagnosis is made by any blood, bone marrow, or stool cultures and with the Widal test (demonstration of antibodies against *Salmonella* antigens O-somatic and H-flagellar). In epidemics and less wealthy countries, after excluding malaria, dysentery, or pneumonia, a therapeutic trial time with chloramphenicol is generally undertaken while awaiting the results of the Widal test and cultures of the blood and stool.

The Widal test is time-consuming and prone to significant false positive results. The test may also be falsely negative in the early course of illness. However, unlike the Typhidot test, the Widal test quantifies the specimen with titres.

Typhidot is a medical test consisting of a dot ELISA kit that detects IgM and IgG antibodies against the outer membrane protein (OMP) of the *Salmonella enterica* subsp. *enterica*. The typhidot test becomes positive within 2–3 days of infection and separately identifies IgM and IgG antibodies. The test is based on the presence of specific IgM and IgG antibodies to a specific 50Kd OMP antigen, which is impregnated on nitrocellulose strips. IgM shows recent infection whereas IgG signifies remote infection. The most important limitation of this test is that it is not quantitative and the result is only positive or negative.

Prevention

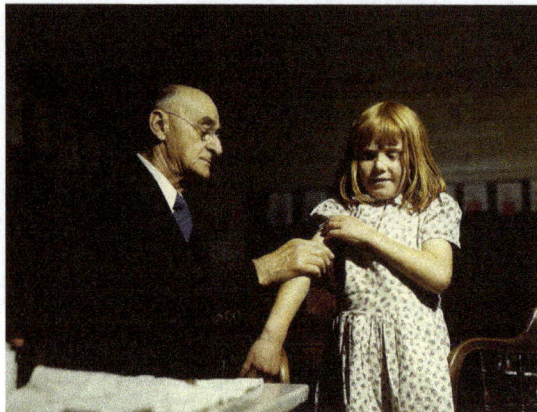

Doctor administering a typhoid vaccination at a school in San Augustine County, Texas, 1943.

Sanitation and hygiene are important to prevent typhoid. It can only spread in environments where human feces are able to come into contact with food or drinking water. Careful food preparation and washing of hands are crucial to prevent typhoid. Industrialization, and in particular, the invention of the automobile, contributed greatly to the elimination of typhoid fever, as it eliminated the public-health hazards associated with having horse manure in public streets, which led to large number of flies, which are known as vectors of many pathogens, including *Salmonella*spp. According to statistics from the United States Centers for Disease Control and Prevention, the chlorination of drinking water has led to dramatic decreases in the transmission of typhoid fever in the United States.

Vaccination

Two typhoid vaccines are licensed for use for the prevention of typhoid: the live, oral Ty21a vaccine (sold as Vivotif by Crucell Switzerland AG) and the injectable typhoid polysaccharide vaccine (sold as Typhim Vi by Sanofi Pasteur and Typherix by GlaxoSmithKline). Both are efficacious and recommended for travellers to areas where typhoid is endemic. Boosters are recommended every five years for the oral vaccine and every two years for the injectable form. An older, killed whole-cell vaccine is still used in countries where the newer preparations are not available, but this vaccine is no longer recommended for use because it has a higher rate of side effects (mainly pain and inflammation at the site of the injection).

To help decrease rates of typhoid fever in developing nations, the World Health Organization (WHO) endorsed the use of a vaccination program starting in 1999. Vaccinations have proven to be a great way at controlling outbreaks in high incidence areas. Just as important, it is also very cost-effective. Vaccination prices are normally low, less than US $1 per dose. Because the price is low, poverty-stricken communities are more willing to take advantage of the vaccinations. Although vaccination programs for typhoid have proven to be effective, they alone cannot eliminate typhoid fever. Combining the use of vaccines with increasing public health efforts is the only proven way to control this disease.

Since the 1990s, two typhoid fever vaccines have been recommended by the WHO. The ViPS vaccine is given via injection, while the Ty21a is taken through capsules. Only people 2 years or older are recommended to be vaccinated with the ViPS vaccine, and it requires a revaccination after 2–3 years with a 55–72% vaccine efficacy. The alternative Ty21a vaccine is recommended for people 5 years or older, and has a 5-7-year duration with a 51–67% vaccine efficacy. The two different vaccines have been proven as a safe and effective treatment for epidemic disease control in multiple regions.

A version combined with hepatitis A is also available.

Treatment

Oral Rehydration Therapy

The rediscovery of oral rehydration therapy in the 1960s provided a simple way to prevent many of the deaths of diarrheal diseases in general.

Antibiotics

Where resistance is uncommon, the treatment of choice is a fluoroquinolone such as ciprofloxacin. Otherwise, a third-generation cephalosporin such as ceftriaxone or cefotaxime is the first choice. Cefixime is a suitable oral alternative.

Typhoid fever, when properly treated, is not fatal in most cases. Antibiotics, such as ampicillin, chloramphenicol, trimethoprim-sulfamethoxazole, amoxicillin, and ciprofloxacin, have been commonly used to treat typhoid fever. Treatment of the disease with antibiotics reduces the case-fatality rate to about 1%.

Without treatment, some patients develop sustained fever, bradycardia, hepatosplenomegaly, abdominal symptoms, and occasionally, pneumonia. In white-skinned patients, pink spots, which fade on pressure, appear on the skin of the trunk in up to 20% of cases. In the third week, untreated cases may develop gastrointestinal and cerebral complications, which may prove fatal in up to 10–20% of cases. The highest case fatality rates are reported in children under 4 years. Around 2–5% of those who contract typhoid fever become chronic carriers, as bacteria persist in the biliary tract after symptoms have resolved.

Surgery

Surgery is usually indicated if intestinal perforation occurs. One study found a 30-day mortality rate of 9.1% (8/88), and surgical site infections at 67.0% (59/88).

For surgical treatment, most surgeons prefer simple closure of the perforation with drainage of the peritoneum. Small-bowel resection is indicated for patients with multiple perforations. If antibiotic treatment fails to eradicate the hepatobiliary carriage, the gallbladder should be resected. Cholecystectomy is not always successful in eradicating the carrier state because of persisting hepatic infection.

Resistance

As resistance to ampicillin, chloramphenicol, trimethoprim-sulfamethoxazole, and streptomycin is now common, these agents have not been used as first–line treatment of typhoid fever for almost 20 years. Typhoid resistant to these agents is known as multidrug-resistant typhoid.

Ciprofloxacin resistance is an increasing problem, especially in the Indian subcontinent and Southeast Asia. Many centres are shifting from using ciprofloxacin as the first line for treating suspected typhoid originating in South America, India, Pakistan, Bangladesh, Thailand, or Vietnam. For these people, the recommended first-line treatment is ceftriaxone. Also, azithromycin has been suggested to be better at treating resistant typhoid in populations than both fluoroquinolone drugs and ceftriaxone. Azithromycin significantly reduces relapse rates compared with ceftriaxone.

A separate problem exists with laboratory testing for reduced susceptibility to ciprofloxacin; current recommendations are that isolates should be tested simultaneously against ciprofloxacin (CIP) and against nalidixic acid (NAL), and that isolates that are sensitive to both CIP and NAL should be reported as "sensitive to ciprofloxacin", but that isolates testing sensitive to CIP but not to NAL should be reported as "reduced sensitivity to ciprofloxacin". However, an analysis of 271 isolates showed that around 18% of isolates with a reduced susceptibility to ciprofloxacin (MIC 0.125–1.0 mg/l) would not be picked up by this method. How this problem can be solved is not certain, because most laboratories around the world (including the West) are dependent on disk testing and cannot test for MICs.

Other Water Diseases due to Water Pollution

Waterborne diseases are a result of contaminated water. Pathogenic microorganisms thrive and reproduce in water sources. Bacteria, protozoa, viruses, and intestinal parasites are the most common microorganisms that cause waterborne diseases. Waterborne diseases lead to symptoms that are unpleasant to both body and mind. For some of them, the treatment is easy. But in certain cases, if no form of sufficient medical treatment is provided it can be extremely fatal.

Malarial Fever

The plasmodium parasite mosquito is responsible for the spread of malaria. Water bodies such as lakes and stagnant water are the breeding areas of these harmful parasites. Fevers, headaches, body chills and vomiting are the symptoms of malaria.

Amebiasis

Protozoa is the microorganism that causes amebiasis. The most common symptoms are diarrhoea, fatigue, abdominal discomfort, flatulence, and weight loss.

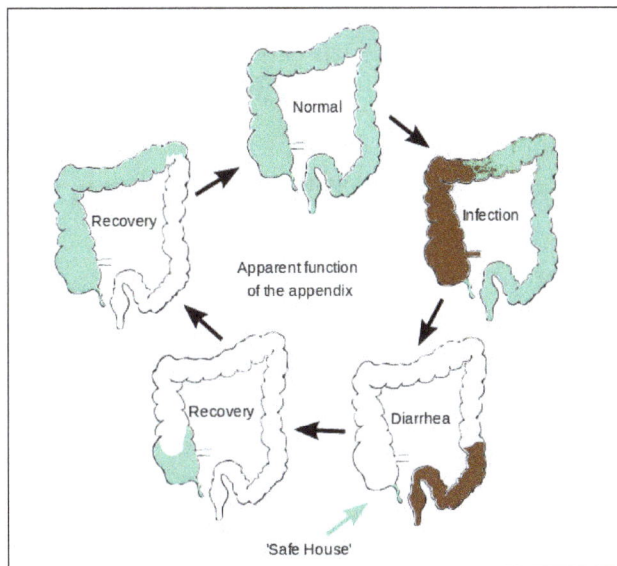

Hepatitis

This waterborne disease is caused by a virus. Fever, jaundice, chills, abdominal discomfort and dark urine are the most common symptoms.

Viral Gastroenteritis

This stomach infection is contracted when you consume contaminated food or water or come in contact with an infected person. The symptoms of this viral infection are vomiting, gastrointestinal discomfort, headache, and fever.

Cryptosporidiosis

This disease is caused by protozoa and leads to both gastrointestinal and respiratory illness. Watery diarrhoea with or without repeated coughing are the major symptoms.

Filariasis

People who reside near sewages or unsanitary water bodies are affected by filariasis. The infection is spread mostly spread by mosquitoes. They're the host for the harmful filarial nematode worm that affects humans and further leads to elephantiasis.

Shigellosis

The bacteria Shigella is responsible for the infectious disease Shigellosis. People who are infected display symptoms such as fever, diarrhoea, and stomach cramps.

How do you Protect yourself from Waterborne Diseases?

- Never be under the impression that bottled water is always safe. Ensure that your bottle of water is always placed in a dry place. Keep it at room temperature or cooler and away from direct sunlight.

- If you suffer from a weak immune system, you need to ensure that the source of your drinking water is safe and pure. Individuals who are old, undergone an organ transplant or suffer from a chronic disease need to be extra careful.

- If you're not certain about the purity of your drinking water, make sure to boil your water regularly. Simultaneously, as an alternative, you could get in touch with a trusted vendor for bottled water.

- You need to make sure that your water supply is not exposed to any kind pesticides or harsh chemicals. These harmful substances may lead to waterborne infections and other physical ailments.

- Install a high-quality water purifier that'll get rid of the harmful microorganisms and harsh metals that are present in untreated water. This step will ensure that you have access to purified water every single day.

Waterborne diseases can cause serious physical ailments and if not treated properly or in time can lead to death as well. It's important you pay attention to your daily water intake and consume only purified water.

References

- Diseases, water: environmentalpollutioncenters.org, Retrieved 26 July, 2017

- "Ceiba pentandra". Human Uses and Cultural Importance. Archived from the original on 15 February 2012. Retrieved 7 February 2012

- Syc-20355287, symptoms-causes, cholera, diseases-conditions: mayoclinic.org, Retrieved 27 August, 2019

- Health, NPS: Better choices, Better. "Vivaxim Solution for injection". NPS medicinewise. Archived from the originalon 1 October 2015. Retrieved 10 April 2017

- Common-waterborne-diseases-and-how-to-prevent-them: livpure.com, Retrieved 28 January, 2019

6

Prevention and Control of Water Pollution

Some of the ways that can be used to control and prevent water pollution are management of industrial waste, avoiding hazardous material and disposal of toxic wastes, cleaning of drains, recycling or reusing water. The aim of this chapter is to explore these various methods for prevention and control of water pollution.

Prevention from Water Pollution

This list encompasses a lot of things you can do in your own home and yard, but it also gives some suggestions for how to prevent water pollution from industries as well. No matter how you are looking to get involved in the world of water pollution prevention, you can easily find the right way to make some changes by following one or more of the tips on this list.

Do not Dump in or Around Rivers

Dumping is one of the leading causes of water pollution, and you can keep freshwater sources from becoming contaminated by refraining from dumping in them. No matter what you might be dumping, whether it's solid or liquid waste or even something you feel might not be all that harmful, you'll be introducing a new substance into the water supply. This automatically contaminates it and starts it on a fast path toward full pollution. Keep trash and waste out of rivers in order to keep drinking water sources safe.

Clean up Rivers that have a Lot of Trash in and around them

If you notice a lot of dumping going on in and around rivers in your area, it's not too late to prevent full-on pollution of these water sources. Get involved and start cleaning them up right away to have the best possible chance of preventing long-term effects from setting in. If you see litter on the banks of rivers, stop and pick it up. A few pieces of litter may soon escalate into a lot of dumping that can't be reversed so easily, but you can do a lot to prevent pollution by picking up trash when you see it, whether it's big or small.

Clean and Maintain Septic Systems Appropriately

This may not be entirely up to you, depending on your city or county, but if you have a septic system try to do your part to keep it as clean and well-maintained as possible. Septic systems have a tendency to get too dirty or to start leaking, and when this happens, human waste is exposed to groundwater almost immediately. If you live near freshwater sources, this can also further pollute the surface water easily. You can prevent human waste pollution in water by making sure your septic system is fully functional and in good shape at all times. Call for maintenance checks at least twice a year.

Follow all Water Laws and Regulations given out by your Municipality

These regulations and laws are in place for a reason, and they're usually there to protect your water supplies. When you break these rules, you're putting the water at risk by participating in activities that are frowned upon or sometimes completely illegal. If you know of water regulations in your area, be sure to keep them in mind when it comes time for you to do anything around your home such as treat your lawn, wash your car, or even water your plants or grass.

Talk to your City or County about how their Water Supplies are being Treated

If you have city or county water instead of a well, you aren't in charge of the way the water is treated for contaminants. However, that doesn't mean you aren't allowed to find out. Regular water quality reports should be made available to the public upon request in most municipalities, and you can also ask for more information about what contaminants and pollutants are being treated for in your water supply. If you're concerned about a specific potential pollution problem, you can contact your water company and find out if they're doing anything to combat it, or if it's something you can take care of in your own home.

Talk to your City or county about Maintaining and Cleaning Sewer Lines Regularly

This is very similar to the problem of septic systems, but in this situation, you aren't able to do much about it if you suspect your sewage system is in need of repair. However, if you believe there is a real potential for contamination from your sewage system—especially if you're noticing a strange smell from your tap or waste is coming up through your water pipes—be sure to get in touch with your local water company right away. They can help you reach the right people to take care of this problem as soon as possible and prevent you and your water from becoming polluted.

Always Dispose of Pesticides and other Harsh Chemicals Appropriately

Never pour these substances out in your yard, and never pour them down storm drains. Do not pour them down the sink or in the toilet, either. Eventually, all of these pipes and drains will lead to water sources, and if they're carrying harsh chemicals, those chemicals will enter into the water supply as well. This is a vicious cycle, and the chemicals are sure to get back around to drinking water eventually. If you empty pesticides and cleaning chemicals into your yard, you run the risk of polluting groundwater that eventually reaches surface water as well. This is also true of medication and pills. Much like pesticides and chemicals, they shouldn't be disposed of down the toilet or sink either.

Never Bury Animals in the Backyard

Although it can be difficult determining what to do with a beloved family pet after it has passed on, burying it in the yard can have dire consequences for the whole family. If you absolutely must bury an animal, be sure to place it in an environmentally-friendly box that won't allow bacteria to escape into your soil and potentially reach your groundwater. The best way to handle the body of a deceased family pet is to contact your veterinarian or even animal control to discuss what to do.

Never Dispose of cooking Fats and Oils by Pouring them down the Sink

Not only can this seriously clog up your drains, but it can also cause disease and illness to spread in your local water supply. This is a big problem with groundwater contamination in yards, so if you have a well you should be doubly concerned with properly disposing of fat, grease, and oil. If these substances get stuck in pipes, they can leach into the surrounding soil easily. They attract bacteria almost right away, and those bacteria contribute to the spread of disease in water sources.

Set up a Composting Pile and do not use a Garbage Disposal Unless you absolutely have to

Although many homes are fitted with garbage disposals, they can cause a buildup of bacteria in groundwater around your home much like disposing of fats and oils down the drain can. This is a big problem in largely residential areas, but you can do your part to prevent this type of pollution by throwing solid waste on a composting pile and using it in your yard or home garden instead. By recycling this waste instead of just chopping it up and washing it down the drain, you're doing a lot for the environment, and you're also keeping unwanted materials out of water supplies further down the line.

Never Pour Oil or Fuel down the Drain or into Storm Drains

These substances can cause a lot of trouble when disposed of incorrectly. Not only can they clog up your water pipes like cooking fats and oils can, but they can also eat away at those pipes and eventually cause serious leaks underground. These leaks can then contaminate the groundwater in your area and quickly cause pollution in surface water sources as well. Always dispose of fuel and oil properly, and take care not to work on your vehicles in your yard to keep from accidentally spilling these substances onto the ground or down storm drains.

Refrain from using Bleach when Washing Dishes and Laundry if at all Possible

Sometimes bleach is necessary, so it may not be possible to completely cut it out of your life. However, only use it when you absolutely have to in order to prevent contamination in nearby water supplies. When bleach is present in wastewater, it is washed into water sources that eventually lead to drinking water supplies. Bleach can be very damaging to humans, animals, and plant life when it's present in large enough quantities in water sources. It can cause internal burning as well as poisoning, and it can seriously damage the environment surrounding water sources.

Encourage Factories, Construction Sites, and Agricultural Sites in your area to use Safe Runoff Practices

Toxic runoff is present in all of these locations, but it doesn't have to be a huge problem if these industries take care to prevent it from reaching surface water sources. Groundwater pollution is a huge problem that absolutely must be taken into consideration as well, but runoff may also reach fresh surface water, especially when these industries are near sources of fresh water. The more environmentally friendly all of these sites are, the better off the water in that area will be. It's impossible to prevent industries from operating in any area, but it isn't impossible for those industries to prevent water pollution at one of its biggest sources.

Clean up your Yard Regularly, and Encourage your Neighbors to do the same

Don't let garbage items like cigarette butts or stray pieces of trash sit around in your yard for too long. Some neighborhoods help to eliminate this problem, but others don't, and it can sometimes be a huge issue. If you notice your neighborhood getting very dirty over time, organize a cleanup day and get people in the area involved. This is a great community effort to help cut back on water pollution.

Work to Educate your Friends, Family, and the Community about Water Safety and Water Pollution

Although you might be a little nervous about bringing it up with people in your area, chances are you can make a big difference in the prevention of water pollution by simply educating the people around you about this very real problem. Some of them might not know about the small steps they can take to prevent water pollution, and even some industries might be more inclined to work toward prevention of this problem if they know there are concerned people living in the areas where they operate. Don't be afraid to reach out to people in governmental positions in your county or city and let them know about the ways you'd like to see water pollution prevention taking place, too.

Ways to Reduce Water Pollution

Highly contaminated water coming out from the industries and their leftover chemical residues, etc are also discharged into the river through drains. Waste generated, due to daily activities of the people living in the houses, are also thrown into the rivers which leaves the river's water highly polluted. If we have to control water pollution, we will have to find out a way out, and devise laws and strategies.

Enforcing Laws to Prevent Water Pollution

We should Strictly Follow all the Laws Regarding Water Pollution

The legislative provisions, such as the Water Act 1974 and Control of Pollution Prevention and Environmental Protection Act 1986 are there but these have not been implemented effectively and so we will have to get these implemented strictly for effective prevention of water pollution. Water Cess Act 1977 is another important law which aims to reduce and prevent water pollution; however, its effects have been limited. Apart from the laws, creating awareness about the impacts of water pollution is required. Through public awareness and effective implementation of established laws, water pollution can be reduced very effectively.

Industries should Behave more Responsibly

Many industries directly flow their waste everywhere which reaches rivers through rain water. To prevent water pollution from industrial wastes, it is required that these wastes should be disposed of properly. Some industries follow this rule, and they either destroy the remaining material, or re-use it safely. In addition to applying these methods, industries are required to bring about changes into their methods of manufacturing to prevent water pollution. But not all the industries are following these norms. Most of them throw their wastes into the rivers which is a dangerous scenario as far as water pollution is concerned, as all these wastes finally affected the water animals as well as the humans.

Avoiding Hazardous Material

It is also extremely important to adopt the correct methods of the disposal of toxic wastes. In the places where paints, cleaning and stain removal chemicals are used, it is required to arrange for the safe disposal of the wastes and the contaminated water coming out of these factories. Oil spill from cars, other vehicles and the machines are required to be stopped completely.

Oil leak from the cars and other machines have posed bigger threat and these have become major contributors for water pollution. So, it is important to take care of cars and machines. Oil leak from the factories are also required to be stopped after the completion of the work in factories. These factories are required to apply all the ways for the safe disposal and clearance of the oil.

Cleaning of Drains

To prevent water pollution, the drains are required to be cleaned on a regular basis. In the rural areas, pucca drains are required to be made, because the water is going everywhere in a chaotic manner; it finally reaches the rivers and canals with tons of garbage and pollutants. We should develop a technology to keep the drains away from the water sources.

Recycling and Reuse of Water

Re-cycling and re-use are other ways to prevent water pollution which can improve the availability of fresh water. The use of low quality water, such as treated wastewater in the industries and for washing utensils and gardening makes the fresh water less contaminated. Such water can also be used for washing vehicles and we should use only good quality water for drinking purposes. Currently, water recycling is being only in a limited manner. So, we will have to stress more on proper recycling and reuse of water to prevent water pollution.

Preventing Soil Erosion

To prevent water from getting polluted, we are also required to prevent soil erosion. If there is soil conservation, we can stop water pollution up to some extent. We will have to plant more trees to stop soil erosion. We must adopt such methods which can cultivate the soil and improve the health of the environment.

Cleaning of Water ways and the Beaches

Cleaning is required on a regular basis as water of the rivers, ponds and even the groundwater has also been contaminated by humans. Even the humans have not spared the ocean water, making it polluted. Travel through sea, growing preferences for residing near sea shores has resulted into several small and big settlements near seas which has made the pollution of sea water a rising concern. For their livelihood many people are selling different contents to the tourists and they throw the residue at sea shores and thus the water of the sea become polluted.

The temporary settlements near the sea normally do not have toilets which is why these people defecate in the water of the ocean and the people also throw their household garbage in ocean waters. After cleaning up their mess the ships also throw their garbage into the water. Sometimes, accidents of the ships also takes place into the sea and thus various chemical substances and oil get spilled over the sea water leaving long term impact on the creatures living in water.

Due to contamination of water some organisms die immediately and make water more polluted. The consumption of these aquatic organisms living in polluted sea water also makes the humans go sick. Developed countries also throw their e-wastes and poisonous garbage into the sea and thus sea water gets badly contaminated.

Need for Living in Harmony with Nature

Man has forgotten that his existence on this Earth is because of the nature and environment. The human negligence is also a major cause of environmental pollution. Various species and organisms of the water naturally die due to humans bathing in the water and thus making it polluted. Household wastes and industrial waste also add to the problem. It is time for learning sustainable ways of living.

Adopting Organic Farming

The farmers should stop using various chemical fertilizers in their fields to get a bumper harvest or spraying pesticides on their crops for this purpose. When it rains all chemicals goes into the ponds and rivers through rain water and thus water bodies get heavily polluted.

Best Management Practice for Water Pollution

Best management practices (BMPs) is a term used in the United States and Canada to describe a type of water pollution control. Historically the term has referred to auxiliary pollution controls in the fields of industrial wastewater control and municipal sewage control, while in stormwater management (both urban and rural) and wetland management, BMPs may refer to a principal control or treatment technique as well.

A retention pond for treatment of urban runoff (stormwater).

Beginning in the 20th century, designers of industrial and municipal sewage pollution controls typically utilized engineered systems (e.g. filters, clarifiers, biological reactors) to provide the central components of pollution control systems, and used the term "BMPs" to describe the supporting functions for these systems, such as operator training and equipment maintenance.

Stormwater management, as a specialized area within the field of environmental engineering, emerged later in the 20th century, and some practitioners have used the term BMP to describe both structural or engineered control devices and systems (e.g. retention ponds) to treat polluted stormwater, as well as operational or procedural practices (e.g. minimizing use of chemical fertilizers and pesticides). Other practitioners prefer to use the term Stormwater control measure, due to the varied definitions of the term "BMP" and its use in non-stormwater practice.

U.S. Clean Water Act References to "BMP"

Congress referred to BMP in several sections of the U.S. Clean Water Act (CWA) but did not define the term.

- The 1977 CWA used the term in describing the areawide waste treatment planning program and in procedures for controlling toxic pollutants associated with industrial discharges. The "Section 404" program, which covers dredge and fill permits, refers to BMPs in one of the enforcement exemptions.

- References to stormwater BMPs first appear in the 1987 amendment to the CWA in describing the Nonpoint Source Management Demonstration Program.

- Another stormwater BMP reference was added in 2001 with the authorization for a Wet Weather Watershed Pilot Project program.

EPA Definitions

In implementing the CWA, the U.S. Environmental Protection Agency (EPA) defined BMP in the federal wastewater permit regulations, initially to refer to auxiliary procedures for industrial wastewater controls.

> "Schedules of activities, prohibitions of practices, maintenance procedures, and other management practices to prevent or reduce the pollution of waters of the United States, BMPs also include treatment requirements, operating procedures, and practices to control plant site runoff, spillage or leaks, sludge or waste disposal, or drainage from raw material storage."

Later the Agency added a reference to stormwater management BMPs.

> "Each NPDES permit shall include conditions meeting the following requirements when applicable (k) Best management practices (BMPs) to control or abate the discharge of pollutants when: Authorized under section 402(p) of the CWA for the control of storm water discharges."

Industrial Wastewater BMPs

Industrial wastewater BMPs are considered an adjunct to engineered treatment systems. Typical BMPs include operator training, maintenance practices, and spill control procedures for treatment chemicals. There are also many BMPs available which are specific to particular industrial processes, for example:

- Source reduction practices in metal finishing industries (e.g. substituting less toxic solvents or using water-based cleaners);

- In the chemical industry, capturing equipment washdown waters for recycle/reuse at various process stages;

- In the paper industry, using process control monitoring to optimize bleaching processes, and reduce the overall amount of bleach used.

Stormwater Management BMPs

Stormwater management BMPs are control measures taken to mitigate changes to both quantity and quality of urban runoff caused through changes to land use. Generally BMPs focus on water quality problems caused by increased impervious surfaces from land development. BMPs are designed to reduce stormwater volume, peak flows, and/or nonpoint source pollution through evapotranspiration, infiltration, detention, and filtration or biological and chemical actions. BMPs also can improve receiving-water quality by extending the duration of outflows in comparison to inflow duration (known as hydrograph extension), which dilutes the stormwater discharged into a larger volume of upstream flow.

Stormwater BMPs can be classified as "structural" (i.e., devices installed or constructed on a site like Sediment Control Fence, Rock Filter Dams, Erosion Control Logs, Excelsior Wattle, Sediment Traps and numerous other proprietary products) or "non-structural" (procedures, such as modified landscaping practices, soil disturbing activity scheduling, or street sweeping). There are a variety of BMPs available; selection typically depends on site characteristics and pollutant removal objectives. EPA has published a series of stormwater BMP fact sheets for use by local governments, builders and property owners.

Stormwater management BMPs can be also categorized into four basic types:

1. Storage practices: ponds; recovery; green infrastructure design.

2. Vegetative practices: buffers; channels; green roofs; wetlands; functional art; stormwater wetland park design; wetland park engineering & design.

3. Filtration/Infiltration practices: filtering; infiltration; rain gardens; porous pavement; civic infrastructure and design; functional stormwater design.

4. Water sensitive development: better site design; open space site design; low impact development.

7

Water Treatment

The process used to improve the quality of water to make it potable for various uses is referred to as water treatment. The most common methods that are used for water treatment are sedimentation, filtration, aeration and chlorination. This chapter has been carefully written to provide an easy understanding of these methods of water treatment.

Water pollution occurs when undesirable effluents disperse in a water system and so water quality change. Water pollution divided into three main sources, natural sources: include thermal and acid effluents from volcanic areas and are not common on the earth, domestic sources that are primarily sewage and laundry wastes and generated in houses, apartments, and other dwellings. In rural and some suburban areas, domestic wastes are handled at the individual residence and enter the environment through the soil either in partially treated or untreated fashion. In urban areas, domestic wastes are collocated in sewage pipes and transmitted to control location either for treatment or discharge into a watercourse without treatment (This considered as the major potential source of water pollution). Urban sewage since they handled by established government agencies, they can usually be effectively controlled. Industrial wastes vary from industry to industry and from location to location. Some industries generate wastes high in organic matter, and these wastes can usually handle by methods similar to those used for domestic wastes, such industries include dairy and food-processing plants, meat-packing houses. Other industries, however, generate wastes that are low in organic matter but high in toxic chemicals such as metals, acids or alkalis. These include chemical plants, mining facilities, and textile mills.

Dalecarlia Water Treatment Plant, Washington, D.C.

Water treatment is any process that improves the quality of water to make it more acceptable for a specific end-use. The end use may be drinking, industrial water supply, irrigation, river flow maintenance, water recreation or many other uses, including being safely returned to the environment.

Water treatment removes contaminants and undesirable components, or reduces their concentration so that the water becomes fit for its desired end-use. This treatment is crucial to human health and allows humans to benefit from both drinking and irrigation use.

Drinking Water Treatment

Treatment for drinking water production involves the removal of contaminants from raw water to produce water that is pure enough for human consumption without any short term or long term risk of any adverse health effect. In general terms, the greatest microbial risks are associated with ingestion of water that is contaminated with human or animal (including bird) faeces. Faeces can be a source of pathogenic bacteria, viruses, protozoa and helminths. Substances that are removed during the process of drinking water treatment, Disinfection is of unquestionable importance in the supply of safe drinking-water. The destruction of microbial pathogens is essential and very commonly involves the use of reactive chemical agents such suspended solids, bacteria, algae, viruses, fungi, and minerals such as iron and manganese. These substances continue to cause great harm to several lower developed countries that do not have access to water purification.

Measures taken to ensure water quality not only relate to the treatment of the water, but to its conveyance and distribution after treatment. It is therefore common practice to keep residual disinfectants in the treated water to kill bacteriological contamination during distribution.

Water supplied to domestic properties, for tap water or other uses, may be further treated before use, often using an in-line treatment process. Such treatments can include water softening or ion exchange. Many proprietary systems also claim to remove residual disinfectants and heavy metal ions.

Processes

Empty aeration tank for iron precipitation.

The processes involved in removing the contaminants include physical processes such as settling and filtration, chemical processes such as disinfection and coagulation and biological processes such as slow sand filtration.

A combination selected from the following processes is used for municipal drinking water treatment worldwide.

Chlorination

Water chlorination is the process of adding chlorine or chlorine compounds such as sodium hypochlorite to water. This method is used to kill certain bacteria and other microbes in tap water as chlorine is highly toxic. In particular, chlorination is used to prevent the spread of waterborne diseases such as cholera, dysentery, and typhoid.

Checking chlorine level of the local water source in La Paz, Honduras.

In a paper published in 1894, it was formally proposed to add chlorine to water to render it "germ-free". Two other authorities endorsed this proposal and published it in many other papers in 1895. Early attempts at implementing water chlorination at a water treatment plant were made in 1893 in Hamburg, Germany. In 1897 the town of Maidstone, England was the first to have its entire water supply treated with chlorine.

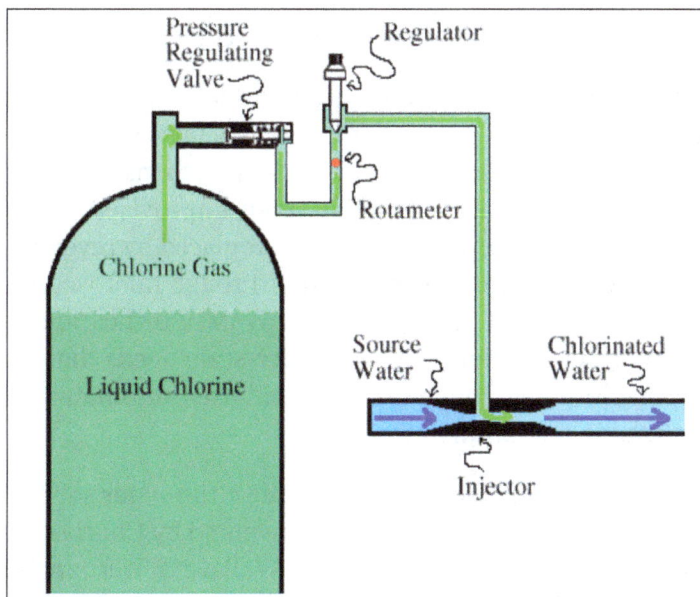
Chlorination.

Permanent water chlorination began in 1905, when a faulty slow sand filter and a contaminated water supply caused a serious typhoid fever epidemic in Lincoln, England. Alexander Cruickshank Houston used chlorination of the water to stop the epidemic. His installation fed a concentrated solution of so-called *chloride of lime* to the water being treated. This was not simply modern calcium chloride, but contained chlorine gas dissolved in lime-water (dilute calcium hydroxide) to form calcium hypo-chlorite (chlorinated lime). The chlorination of the water supply helped stop the epidemic and as a precaution, the chlorination was continued until 1911 when a new water supply was instituted.

The first continuous use of chlorine in the United States for disinfection took place in 1908 at Boonton Reservoir (on the Rockaway River), which served as the supply for Jersey City, New Jersey. Chlorination was achieved by controlled additions of dilute solutions of chloride of lime (calcium hypochlorite) at doses of 0.2 to 0.35 ppm. The treatment process was conceived by John L. Leal, and the chlorination plant was designed by George Warren Fuller. Over the next few years, chlorine disinfection using chlorides of lime (calcium hypochlorite) were rapidly installed in drinking water systems around the world.

The technique of purification of drinking water by use of compressed liquefied chlorine gas was developed by a British officer in the Indian Medical Service, Vincent B. Nesfield, in 1903. According to his own account, "It occurred to me that chlorine gas might be found satisfactory, if suitable means could be found for using it. The next important question was how to render the gas portable. This might be accomplished in two ways: By liquefying it and storing it in lead-lined iron vessels, having a jet with a very fine capillary canal, and fitted with a tap or a screw cap. The tap is turned on, and the cylinder placed in the amount of water required. The chlorine bubbles out, and in ten to fifteen minutes the water is absolutely safe. This method would be of use on a large scale, as for service water carts."

Major Carl Rogers Darnall, Professor of Chemistry at the Army Medical School, gave the first practical demonstration of this in 1910. This work became the basis for present day systems of municipal water *purification*. Shortly after Darnall's demonstration, Major William J. L. Lyster of the Army Medical Department used a solution of calcium hypochlorite in a linen bag to treat water.

For many decades, Lyster's method remained the standard for U.S. ground forces in the field and in camps, implemented in the form of the familiar Lyster Bag (also spelled Lister Bag). The canvas "bag, water, sterilizing" was a common component of field kitchens, issued one per 100 persons, of a standard 36-gallon capacity that hung from an often-improvised tripod in the field. In use from World War I through the Vietnam War, it has been replaced by reverse osmosis systems that produce potable water by pressure straining local water through microscopic-level filters: the Reverse Osmosis Water Purification Unit and the Tactical Water Purification System for large-scale production, and the Light Water Purifier unit for smaller-scale needs that includes ultrafiltration technology to produce potable water from any source and uses automated backwash cycles every 15 minutes to simplify cleaning operations.

Chlorine gas was first used on a continuing basis to disinfect the water supply at the Belmont filter plant, Philadelphia, Pennsylvania by using a machine invented by Charles Frederick Wallace who dubbed it the Chlorinator. It was manufactured by the Wallace & Tiernan Company beginning in 1913. By 1941, disinfection of U.S. drinking water by chlorine gas had largely replaced the use of chloride of lime.

Biochemistry

As a halogen, chlorine is a highly efficient disinfectant, and is added to public water supplies to kill disease-causing pathogens, such as bacteria, viruses, and protozoans, that commonly grow in water supply reservoirs, on the walls of water mains and in storage tanks. The microscopic agents of many diseases such as cholera, typhoid fever, and dysentery killed countless people annually before disinfection methods were employed routinely.

Chlorine is manufactured from salt by electrolysis or other methods. It is a gas at atmospheric pressures but liquid at high pressure. The liquefied gas is transported and used as such.

As a strong oxidizing agent, chlorine kills via the oxidation of organic molecules. Chlorine and hydrolysis product hypochlorous acid are neutrally charged and therefore easily penetrate the negatively charged surface of pathogens. It is able to disintegrate the lipids that compose the cell wall and react with intracellular enzymes and proteins, making them nonfunctional. Microorganisms then either die or are no longer able to multiply.

Principles

When dissolved in water, chlorine converts to an equilibrium mixture of chlorine, hypochlorous acid (HOCl), and hydrochloric acid (HCl):

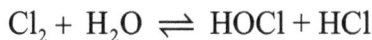

$$Cl_2 + H_2O \rightleftharpoons HOCl + HCl$$

In acidic solution, the major species are Cl_2 and HOCl, whereas in alkaline solution, effectively only ClO^- (hypochlorite ion) is present. Very small concentrations of ClO_2^-, ClO_3^-, ClO_4^- are also found.

Shock Chlorination

Shock chlorination is a process used in many swimming pools, water wells, springs, and other water sources to reduce the bacterial and algal residue in the water. Shock chlorination is performed by mixing a large amount of hypochlorite into the water. The hypochlorite can be in the form of a powder or a liquid such as chlorine bleach (solution of sodium hypochlorite or calcium hypochlorite in water). Water that is being shock chlorinated should not be swum in or drunk until the sodium hypochlorite count in the water goes down to three parts per million (PPM) or until the calcium hypochlorite count goes down to 0.2 to 0.35 PPM.

Drawbacks

Disinfection by chlorination can be problematic, in some circumstances. Chlorine can react with naturally occurring organic compounds found in the water supply to produce compounds known as disinfection by-products (DBPs). The most common DBPs are trihalomethanes (THMs) and haloacetic acids (HAAs). Trihalomethanes are the main disinfectant by-products created from chlorination with two different types, bromoform and dibromochloromethane, which are mainly responsible for health hazards. Their effects depend strictly on the duration of their exposure to the chemicals and the amount ingested into the body. In high doses, bromoform mainly slows down regular brain activity, which is manifested by symptoms such as sleepiness or sedation. Chronic exposure of both bromoform and dibromochloromethane can cause liver and kidney cancer, as well as heart disease, unconsciousness, or death in high doses. Due to the potential carcinogenicity of these compounds, drinking water regulations across the developed world require regular monitoring of the concentration of these compounds in the distribution systems of municipal water systems. The World Health Organization has stated that "the risks to health from these by-products are extremely small in comparison with the risks associated with inadequate disinfection".

There are also other concerns regarding chlorine, including its volatile nature which causes it to disappear too quickly from the water system, and organoleptic concerns such as taste and odor.

Dechlorinator

A dechlorinator is a chemical additive that removes chlorine or chloramine from water. Where tap water is chlorinated, it should be dechlorinated before use in an aquarium, since chlorine can harm aquatic life in the same way it kills micro-organisms. Chlorine will kill fish, and cause damage to an aquarium's biological filter. Chemicals that serve this function are reducing agents which reduce chlorine species to chloride, which is less harmful to fish.

Some compounds employed in commercial dechlorinators are: sodium thiosulfate, Sodium hydroxymethanesulfonate (used in AmQuel), and Sodium hydroxymethane sulfinic acid (used in Marineland Bio-safe).

Aeration

Aeration (also called aerification or aeriation) is the process by which air is circulated through, mixed with or dissolved in a liquid or substance.

Aeration of Liquids

Methods

Aeration of liquids (usually water) is achieved by:

- Passing the liquid through air by means of fountains, cascades, paddle-wheels or cones.

- Passing air through the liquid by means of the Venturi tube, aeration turbines or compressed air which can be combined with diffusers air stones, as well as fine bubble diffusers, coarse bubble diffusers or linear aeration tubing. Ceramics are suitable for this purpose, often involving dispersion of fine air or gas bubbles through the porous ceramic into a liquid. The smaller the bubble, the more gas is exposed to the liquid increasing the gas transfer efficiency. Diffusers or spargers can also be designed into the system to cause turbulence or mixing if desired.

Porous ceramic diffusers are made by fusing aluminum oxide grains using porcelain bonds to form a strong, uniformly porous and homogeneous structure. The naturally hydrophilic material is easily wetted resulting in the production of fine, uniform bubbles.

On a given volume of air or liquid, the surface area changes proportionally with drop or bubble size, the very surface area where exchange can occur. Utilizing extremely small bubbles or drops increases the rate of gas transfer (aeration) due to the higher contact surface area. The pores which these bubbles pass through are generally micrometre-size.

Uses of Aeration of Liquids

Aerated tap water.

- To smooth (laminate) the flow of tap water at the faucet.

- Production of aerated water or cola for drinking purposes.

- Secondary treatment of sewage or industrial wastewater through use of aerating mixers/ diffusers.

- To increase the oxygen content of water used to house animals, such as aquarium fish or fish farm.

- To increase oxygen content of wort (unfermented beer) or must (unfermented wine) to allow yeast to propagate and begin fermentation.

- To dispel other dissolved gases such as carbon dioxide or chlorine.

- In chemistry, to oxidise a compound dissolved or suspended in water.

- To induce mixing of a body of otherwise still water.

Sedimentation

Sedimentation is the tendency for particles in suspension to settle out of the fluid in which they are entrained and come to rest against a barrier. This is due to their motion through the fluid in response to the forces acting on them: these forces can be due to gravity, centrifugal acceleration, or electromagnetism. In geology, sedimentation is often used as the opposite of erosion, i.e., the terminal end of sediment transport. In that sense, it includes the termination of transport by saltation or true bedload transport. Settling is the falling of suspended particles through the liquid, whereas sedimentation is the termination of the settling process. In estuarine environments, settling can be influenced by the presence or absence of vegetation. Trees such as mangroves are crucial to the attenuation of waves or currents, promoting the settlement of suspended particles.

Sedimentation may pertain to objects of various sizes, ranging from large rocks in flowing water to suspensions of dust and pollen particles to cellular suspensions to solutions of single molecules such as proteins and peptides. Even small molecules supply a sufficiently strong force to produce significant sedimentation.

The term is typically used in geology to describe the deposition of sediment which results in the formation of sedimentary rock, but it is also used in various chemical and environmental fields to describe the motion of often-smaller particles and molecules. This process is also used in the bio-tech industry to separate cells from the culture media.

Experiments

In a sedimentation experiment, the applied force accelerates the particles to a terminal velocity v_{term} at which the applied force is exactly canceled by an opposing drag force. For small enough particles (low Reynolds number), the drag force varies linearly with the terminal velocity, i.e., $F_{drag} = f v_{term}$ (Stokes flow) where f depends only on the properties of the particle and the sur-rounding fluid. Similarly, the applied force generally varies linearly with some coupling constant (denoted here as q) that depends only on the properties of the particle, $F_{app} = q E_{app}$. Hence, it is generally possible to define a sedimentation coefficient $s \overset{def}{=} q / f$ that depends only on the proper-ties of the particle and the surrounding fluid. Thus, measuring s can reveal underlying properties of the particle.

In many cases, the motion of the particles is blocked by a hard boundary; the resulting accumula-tion of particles at the boundary is called a sediment. The concentration of particles at the bound-ary is opposed by the diffusion of the particles.

The sedimentation of a single particle under gravity is described by the Mason–Weaver equation, which has a simple exact solution. The sedimentation coefficient s in this case equals m_b / f, where m_b is the buoyant mass.

The sedimentation of a single particle under centrifugal force is described by the Lamm equation, which likewise has an exact solution. The sedimentation coefficient s also equals m_b / f, where m_b is the buoyant mass. However, the Lamm equation differs from the Mason–Weaver equation be-cause the centrifugal force depends on radius from the origin of rotation, whereas in the Mason–Weaver equation gravity is constant. The Lamm equation also has extra terms, since it pertains to sector-shaped cells, whereas the Mason–Weaver equation is one-dimensional.

Classification of sedimentation:

- Type 1 sedimentation is characterized by particles that settle discretely at a constant set-tling velocity, or by the deposition of Iron-Rich minerals to streamlines down to the point source. They settle as individual particles and do not flocculate or stick to other during settling. Example: sand and grit material.

- Type 2 sedimentation is characterized by particles that flocculate during sedimentation and because of this their size is constantly changing and therefore their settling velocity is changing. Example: alum or iron coagulation.

- Type 3 sedimentation is also known as zone sedimentation. In this process the particles are at a high concentration (greater than 1000 mg/L) such that the particles tend to settle as a mass and a distinct clear zone and sludge zone are present. Zone settling occurs in lime-softening, sedimentation, active sludge sedimentation and sludge thickeners.

Filtration

Filtration is a physical, biological or chemical operation that separates solid matter and fluid from a mixture with a filter medium that has a complex structure through which only the fluid can pass. Solid particles that cannot pass through the filter medium are described as oversize and the fluid that passes through is called the filtrate. Oversize particles may form a filter cake on top of the filter and may also block the filter lattice, preventing the fluid phase from crossing the filter, known as blinding. The size of the largest particles that can successfully pass through a filter is called the effective pore size of that filter. The separation of solid and fluid is imperfect; solids will be contaminated with some fluid and filtrate will contain fine particles (depending on the pore size, filter thickness and biological activity). Filtration occurs both in nature and in engineered systems; there are biological, geological, and industrial forms.

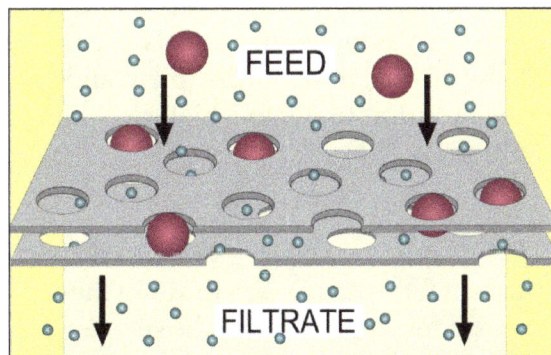

Diagram of simple filtration: oversize particles in the feed cannot pass through the lattice structure of the filter, while fluid and small particles pass through, becoming filtrate.

Process Description

- Filtration is used to separate particles and fluid in a suspension, where the fluid can be a liquid, a gas or a supercritical fluid. Depending on the application, either one or both of the components may be isolated.

- Filtration, as a physical operation is very important in chemistry for the separation of materials of different chemical composition. A solvent is chosen which dissolves one component, while not dissolving the other. By dissolving the mixture in the chosen solvent, one component will go into the solution and pass through the filter, while the other will be retained. This is one of the most important techniques used by chemists to purify compounds.

- Filtration is also important and widely used as one of the unit operations of chemical engineering. It may be simultaneously combined with other unit operations to process the feed stream, as in the biofilter, which is a combined filter and biological digestion device.

- Filtration differs from sieving, where separation occurs at a single perforated layer (a sieve). In sieving, particles that are too big to pass through the holes of the sieve are retained. In filtration, a multilayer lattice retains those particles that are unable to follow the tortuous channels of the filter. Oversize particles may form a cake layer on top of the filter and may also block the filter lattice, preventing the fluid phase from crossing the filter (blinding). Commercially, the term filter is applied to membranes where the separation lattice is so thin that the surface becomes the main zone of particle separation, even though these products might be described as sieves.

- Filtration differs from adsorption, where it is not the physical size of particles that causes separation but the effects of surface charge. Some adsorption devices containing activated charcoal and ion exchange resin are commercially called filters, although filtration is not their principal function.

- Filtration differs from removal of magnetic contaminants from fluids with magnets (typically lubrication oil, coolants and fuel oils), because there is no filter medium. Commercial devices called 'magnetic filters' are sold, but the name reflects their use, not their mode of operation.

- In biological filters, oversize particulates are trapped and ingested and the resulting metabolites may be released. For example, in animals (including humans), renal filtration removes waste from the blood, and in water treatment and sewage treatment, undesirable constituents are removed by absorption into a biological film grown on or in the filter medium, as in slow sand filtration.

Methods of Filtration

There are many different methods of filtration; all aim to attain the separation of substances. Separation is achieved by some form of interaction between the substance or objects to be removed and the filter. The substance that is to pass through the filter must be a fluid, i.e. a liquid or gas. Methods of filtration vary depending on the location of the targeted material, i.e. whether it is dissolved in the fluid phase or suspended as a solid.

Hot Filtration, solution contained in the Erlenmeyer flask is heated on a hot plate in order to prevent re-crystallization of solids in the flask itself.

There are several filtration techniques depending on the desired outcome namely, hot, cold and vacuum filtration. Some of the major purposes of getting the desired outcome are, for the removal of impurities from a mixture or, for the isolation of solids from a mixture.

Hot filtration method is mainly used to separate solids from a hot solution. This is done in order to prevent crystal formation in the filter funnel and other apparatuses that comes in contact with the solution. As a result, the apparatus and the solution used are heated in order to prevent the rapid decrease in temperature which in turn, would lead to the crystallization of the solids in the funnel and hinder the filtration process. One of the most important measures to prevent the formation of crystals in the funnel and to undergo effective hot filtration is the use stemless filter funnel. Due to the absence of stem in the filter funnel, there is a decrease in the surface area of contact between the solution and the stem of the filter funnel, hence preventing re-crystallization of solid in the funnel, adversely affecting filtration process.

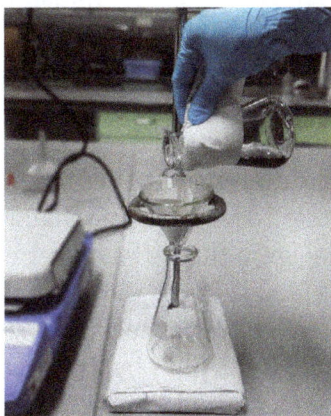

Hot Filtration for the separation of solids from a hot solution.

Cold Filtration method is the use of ice bath in order to rapidly cool down the solution to be crystallized rather than leaving it out to cool it down slowly in the room temperature. This technique results to the formation of very small crystals as opposed to getting large crystals by cooling the solution down at room temperature.

Cold Filtration, the ice bath is used to cool down the temperature of the solution before undergoing the filtration process.

Vacuum Filtration technique is most preferred for small batch of solution in order to quickly dry out small crystals. This method requires a Büchner funnel, filter paper of smaller diameter than the funnel, Büchner flask, and rubber tubing to connect to vacuum source.

Filter Media

Two main types of filter media are employed in laboratories: a surface filter, a solid sieve which traps the solid particles, with or without the aid of filter paper (e.g. Büchner funnel, Belt filter, Rotary vacuum-drum filter, Cross-flow filters, Screen filter); and a depth filter, a bed of granular material which retains the solid particles as it passes (e.g. sand filter). The first type allows the solid particles, i.e. the residue, to be collected intact; the second type does not permit this. However, the second type is less prone to clogging due to the greater surface area where the particles can be trapped. Also, when the solid particles are very fine, it is often cheaper and easier to discard the contaminated granules than to clean the solid sieve.

Filter media can be cleaned by rinsing with solvents or detergents. Alternatively, in engineering applications, such as swimming pool water treatment plants, they may be cleaned by backwashing. Self-cleaning screen filters utilize point-of-suction backwashing to clean the screen without interrupting system flow.

Achieving Flow through the Filter

Fluids flow through a filter due to a difference in pressure—fluid flows from the high-pressure side to the low-pressure side of the filter, leaving some material behind. The simplest method to achieve this is by gravity and can be seen in the coffeemaker example. In the laboratory, pressure in the form of compressed air on the feed side (or vacuum on the filtrate side) may be applied to make the filtration process faster, though this may lead to clogging or the passage of fine particles. Alternatively, the liquid may flow through the filter by the force exerted by a pump, a method commonly used in industry when a reduced filtration time is important. In this case, the filter need not be mounted vertically.

Filter Aid

Certain filter aids may be used to aid filtration. These are often incompressible diatomaceous earth, or kieselguhr, which is composed primarily of silica. Also used are wood cellulose and other inert porous solids such as the cheaper and safer perlite.

These filter aids can be used in two different ways. They can be used as a precoat before the slurry is filtered. This will prevent gelatinous-type solids from plugging the filter medium and also give a clearer filtrate. They can also be added to the slurry before filtration. This increases the porosity of the cake and reduces resistance of the cake during filtration. In a rotary filter, the filter aid may be applied as a precoat; subsequently, thin slices of this layer are sliced off with the cake.

The use of filter aids is usually limited to cases where the cake is discarded or where the precipitate can be chemically separated from the filter.

Alternatives

Filtration is a more efficient method for the separation of mixtures than decantation, but is much more time consuming. If very small amounts of solution are involved, most of the solution may be soaked up by the filter medium.

An alternative to filtration is centrifugation—instead of filtering the mixture of solid and liquid particles, the mixture is centrifuged to force the (usually) denser solid to the bottom, where it often forms a firm cake. The liquid above can then be decanted. This method is especially useful for separating solids which do not filter well, such as gelatinous or fine particles. These solids can clog or pass through the filter, respectively.

Examples of filtration include:

- The coffee filter to keep the coffee separate from the grounds.

- HEPA filters in air conditioning to remove particles from air.

- Belt filters to extract precious metals in mining.

- Horizontal plate filter, also known as Sparkler filter.

- Furnaces use filtration to prevent the furnace elements from fouling with particulates.

- Pneumatic conveying systems often employ filtration to stop or slow the flow of material that is transported, through the use of a baghouse.

- In the laboratory, a Büchner funnel is often used, with a filter paper serving as the porous barrier.

- Air filters are commonly used to remove airborne particulate matter in building ventilation systems, combustion engines, and industrial processes.

- Oil filter in automobiles, often as a canister or cartridge.

- Aquarium filter.

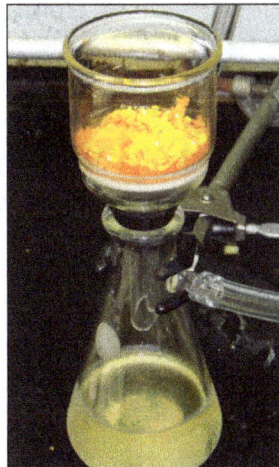

Filter flask (suction flask, with sintered glass filter containing sample).
Note the almost colourless filtrate in the receiver flask.

An experiment to prove the existence of microscopic organisms involves the comparison of water passed through unglazed porcelain and unfiltered water. When left in sealed containers the filtered water takes longer to go foul, demonstrating that very small items (such as bacteria) can be removed from fluids by filtration.

In the kidney, renal filtration is the filtration of blood in the glomerulus, followed by selective re-absorption of many substances essential for the body to maintain homeostasis.

Water Purification

Disinfection is accomplished both by filtering out harmful micro-organisms and by adding disinfectant chemicals. Water is disinfected to kill any pathogens which pass through the filters and to provide a residual dose of disinfectant to kill or inactivate potentially harmful micro-organisms in the storage and distribution systems. Possible pathogens include viruses, bacteria, including *Salmonella*, *Cholera*, *Campylobacter* and *Shigella*, and protozoa, including *Giardia lamblia* and other *cryptosporidia*. After the introduction of any chemical disinfecting agent, the water is usually held in temporary storage – often called a contact tank or clear well – to allow the disinfecting action to complete.

Chlorine Disinfection

The most common disinfection method involves some form of chlorine or its compounds such as chloramine or chlorine dioxide. Chlorine is a strong oxidant that rapidly kills many harmful micro-organisms. Because chlorine is a toxic gas, there is a danger of a release associated with its use. This problem is avoided by the use of sodium hypochlorite, which is a relatively inexpensive solution used in household bleach that releases free chlorine when dissolved in water. Chlorine solutions can be generated on site by electrolyzing common salt solutions. A solid form, calcium hypochlorite, releases chlorine on contact with water. Handling the solid, however, requires more routine human contact through opening bags and pouring than the use of gas cylinders or bleach, which are more easily automated. The generation of liquid sodium hypochlorite is inexpensive and also safer than the use of gas or solid chlorine. Chlorine levels up to 4 milligrams per liter (4 parts per million) are considered safe in drinking water.

All forms of chlorine are widely used, despite their respective drawbacks. One drawback is that chlorine from any source reacts with natural organic compounds in the water to form potentially harmful chemical by-products. These by-products, trihalomethanes (THMs) and haloacetic acids (HAAs), are both carcinogenic in large quantities and are regulated by the United States Environmental Protection Agency (EPA) and the Drinking Water Inspectorate in the UK. The formation of THMs and haloacetic acids may be minimized by effective removal of as many organics from the water as possible prior to chlorine addition. Although chlorine is effective in killing bacteria, it has limited effectiveness against pathogenic protozoa that form cysts in water such as *Giardia lamblia* and *Cryptosporidium*.

Chlorine Dioxide Disinfection

Chlorine dioxide is a faster-acting disinfectant than elemental chlorine. It is relatively rarely used because in some circumstances it may create excessive amounts of chlorite, which is a by-product regulated to low allowable levels in the United States. Chlorine dioxide can be supplied as an aqueous solution and added to water to avoid gas handling problems; chlorine dioxide gas accumulations may spontaneously detonate.

Chloramination

The use of chloramine is becoming more common as a disinfectant. Although chloramine is not as strong an oxidant, it provides a longer-lasting residual than free chlorine because of its lower redox potential compared to free chlorine. It also does not readily form THMs or haloacetic acids (disinfection byproducts).

It is possible to convert chlorine to chloramine by adding ammonia to the water after adding chlorine. The chlorine and ammonia react to form chloramine. Water distribution systems disinfected with chloramines may experience nitrification, as ammonia is a nutrient for bacterial growth, with nitrates being generated as a by-product.

Ozone Disinfection

Ozone disinfection, or ozonation, Ozone is an unstable molecule which readily gives up one atom of oxygen providing a powerful oxidizing agent which is toxic to most waterborne organisms. It is a very strong, broad spectrum disinfectant that is widely used in Europe and in a few municipalities in the United States and Canada. It is an effective method to inactivate harmful protozoa that form cysts. It also works well against almost all other pathogens. Ozone is made by passing oxygen through ultraviolet light or a "cold" electrical discharge. To use ozone as a disinfectant, it must be created on-site and added to the water by bubble contact. Some of the advantages of ozone include the production of fewer dangerous by-products and the absence of taste and odour problems (in comparison to chlorination). No residual ozone is left in the water. In the absence of a residual disinfectant in the water, chlorine or chloramine may be added throughout a distribution system to remove any potential pathogens in the distribution piping.

Ozone has been used in drinking water plants since 1906 where the first industrial ozonation plant was built in Nice, France. The U.S. Food and Drug Administration has accepted ozone as being safe; and it is applied as an anti-microbiological agent for the treatment, storage, and processing of foods. However, although fewer by-products are formed by ozonation, it has been discovered that ozone reacts with bromide ions in water to produce concentrations of the suspected carcinogen bromate. Bromide can be found in fresh water supplies in sufficient concentrations to produce (after ozonation) more than 10 parts per billion (ppb) of bromate — the maximum contaminant level established by the USEPA. Ozone disinfection is also energy intensive.

Ultraviolet Disinfection

Ultraviolet light (UV) is very effective at inactivating cysts, in low turbidity water. UV light's disinfection effectiveness decreases as turbidity increases, a result of the absorption, scattering, and shadowing caused by the suspended solids. The main disadvantage to the use of UV radiation is that, like ozone treatment, it leaves no residual disinfectant in the water; therefore, it is sometimes necessary to add a residual disinfectant after the primary disinfection process. This is often done through the addition of chloramines, consider as a primary disinfectant. When used in this manner, chloramines provide an effective residual disinfectant with very few of the negative effects of chlorination.

Over 2 million people in 28 developing countries use Solar Disinfection for daily drinking water treatment.

Ionizing Radiation

Like UV, ionizing radiation (X-rays, gamma rays, and electron beams) has been used to sterilize water.

Bromination and Iodinization

Bromine and iodine can also be used as disinfectants. However, chlorine in water is over three times more effective as a disinfectant against *Escherichia coli* than an equivalent concentration of bromine, and over six times more effective than an equivalent concentration of iodine. Iodine is commonly used for portable water purification, and bromine is common as a swimming pool disinfectant.

Portable Water Purification

Potable water purification devices and methods are available for disinfection and treatment in emergencies or in remote locations. Disinfection is the primary goal, since aesthetic considerations such as taste, odour, appearance, and trace chemical contamination do not affect the short-term safety of drinking water.

Additional Treatment Options

- Water fluoridation: In many areas fluoride is added to water with the goal of preventing tooth decay. Fluoride is usually added after the disinfection process. In the U.S., fluoridation is usually accomplished by the addition of hexafluorosilicic acid, which decomposes in water, yielding fluoride ions.

- Water conditioning: This is a method of reducing the effects of hard water. In water systems subject to heating hardness salts can be deposited as the decomposition of bicarbonate ions creates carbonate ions that precipitate out of solution. Water with high concentrations of hardness salts can be treated with soda ash (sodium carbonate) which precipitates out the excess salts, through the common-ion effect, producing calcium carbonate of very high purity. The precipitated calcium carbonate is traditionally sold to the manufacturers of toothpaste. Several other methods of industrial and residential water treatment are claimed (without general scientific acceptance) to include the use of magnetic and/or electrical fields reducing the effects of hard water.

- Plumbosolvency reduction: In areas with naturally acidic waters of low conductivity (i.e. surface rainfall in upland mountains of igneous rocks), the water may be capable of dissolving lead from any lead pipes that it is carried in. The addition of small quantities of phosphate ion and increasing the pH slightly both assist in greatly reducing plumbo-solvency by creating insoluble lead salts on the inner surfaces of the pipes.

- Radium Removal: Some groundwater sources contain radium, a radioactive chemical element. Typical sources include many groundwater sources north of the Illinois River in Illinois, United States of America. Radium can be removed by ion exchange, or by water conditioning. The back flush or sludge that is produced is, however, a low-level radioactive waste.

- Fluoride Removal: Although fluoride is added to water in many areas, some areas of the world have excessive levels of natural fluoride in the source water. Excessive levels can be toxic or cause undesirable cosmetic effects such as staining of teeth. A method of reducing fluoride levels is through treatment with activated alumina and bone char filter media.

Other Water Purification Techniques

Other popular methods for purifying water, especially for local private supplies are listed below. In some countries some of these methods are also used for large scale municipal supplies. Particularly important are distillation (de-salination of seawater) and reverse osmosis.

- Boiling: Bringing water to its boiling point (about 100 °C or 212 F at sea level), is the oldest and most effective way since it eliminates most microbes causing intestine related diseases, but it cannot remove chemical toxins or impurities. For human health, complete sterilization of water is not required, since the heat resistant microbes are not intestine affecting. The traditional advice of boiling water for ten minutes is mainly for additional safety, since microbes start getting eliminated at temperatures greater than 60 °C (140 °F). Though the boiling point decreases with increasing altitude, it is not enough to affect the disinfecting process. In areas where the water is "hard" (that is, containing significant dissolved calcium salts), boiling decomposes the bicarbonate ions, resulting in partial precipitation as calcium carbonate. This is the "fur" that builds up on kettle elements, etc., in hard water areas. With the exception of calcium, boiling does not remove solutes of higher boiling point than water and in fact increases their concentration (due to some water being lost as vapour). Boiling does not leave a residual disinfectant in the water. Therefore, water that is boiled and then stored for any length of time may acquire new pathogens.

- Granular Activated Carbon adsorption: a form of activated carbon with a high surface area adsorbs many compounds including many toxic compounds. Water passing through activated carbon is commonly used in municipal regions with organic contamination, taste or odors. Many household water filters and fish tanks use activated carbon filters to further purify the water. Household filters for drinking water sometimes contain silver as metallic silver nanoparticle. If water is held in the carbon block for longer periods, microorganisms can grow inside which results in fouling and contamination. Silver nanoparticles are excellent anti-bacterial material and they can decompose toxic halo-organic compounds such as pesticides into non-toxic organic products. Filtered water must be used soon after it is filtered, as the low amount of remaining microbes may proliferate over time. In general, these home filters remove over 90% of the chlorine available to a glass of treated water. These filters must be periodically replaced otherwise the bacterial content of the water may actually increase due to the growth of bacteria within the filter unit.

- Distillation involves boiling the water to produce water vapour. The vapour contacts a cool surface where it condenses as a liquid. Because the solutes are not normally vaporised, they remain in the boiling solution. Even distillation does not completely purify water, because of contaminants with similar boiling points and droplets of unvapourised liquid carried with the steam. However, 99.9% pure water can be obtained by distillation.

- Reverse osmosis: Mechanical pressure is applied to an impure solution to force pure water through a semi-permeable membrane. Reverse osmosis is theoretically the most thorough method of large scale water purification available, although perfect semi-permeable membranes are difficult to create. Unless membranes are well-maintained, algae and other life forms can colonize the membranes.

- The use of iron in removing arsenic from water.

- Direct contact membrane distillation (DCMD): Applicable to desalination. Heated seawater is passed along the surface of a hydrophobic polymer membrane. Evaporated water passes from the hot side through pores in the membrane into a stream of cold pure water on the other side. The difference in vapour pressure between the hot and cold side helps to push water molecules through.

- Desalination is a process by which saline water (generally sea water) is converted to fresh water. The most common desalination processes are distillation and reverse osmosis. Desalination is currently expensive compared to most alternative sources of water, and only a very small fraction of total human use is satisfied by desalination. It is only economically practical for high-valued uses (such as household and industrial uses) in arid areas.

- Gas hydrate crystals centrifuge method. If carbon dioxide or other low molecular weight gas is mixed with contaminated water at high pressure and low temperature, gas hydrate crystals will form exothermically. Separation of the crystalline hydrate may be performed by centrifuge or sedimentation and decanting. Water can be released from the hydrate crystals by heating.

- In Situ Chemical Oxidation, a form of advanced oxidation processes and advanced oxidation technology, is an environmental remediation technique used for soil and/ or groundwater remediation to reduce the concentrations of targeted environmental contaminants to acceptable levels. ISCO is accomplished by injecting or otherwise introducing strong chemical oxidizers directly into the contaminated medium (soil or groundwater) to destroy chemical contaminants in place. It can be used to remediate a variety of organic compounds, including some that are resistant to natural degradation.

- Bioremediation is a technique that uses microorganisms in order to remove or extract certain waste products from a contaminated area. Since 1991 bioremediation has been a suggested tactic to remove impurities from water such as alkanes, perchlorates, and metals. The treatment of ground and surface water, through bioremediation, with respect to perchlorate and chloride compounds, has seen success as perchlorate compounds are highly soluble making it difficult to remove. Such success by use of *Dechloromonas agitata* strain CKB include field studies conducted in Maryland and the Southwest region of the United States. Although a bioremediation technique may be successful, implementation is not feasible as there is still much to be studied regarding rates and after effects of microbial activity as well as producing a large scale implementation method.

Disinfection

The goal of disinfection of public water supplies is the elimination of the pathogens that are responsible for waterborne diseases. The transmission of diseases such as typhoid and paratyphoid fevers, cholera, salmonellosis, and shigellosis can be controlled with treatments that substantially reduce the total number of viable microorganisms in the water.

While the concentration of organisms in drinking water after effective disinfection may be exceedingly small, sterilization (i.e., killing all the microbes present) is not attempted. Sterilization is not only impractical, it cannot be maintained in the distribution system. Assessment of the reduction in microbes that is sufficient to protect against the transmission of pathogens in water.

Chlorination is the most widely used method for disinfecting water supplies in the United States. The near universal adoption of this method can be attributed to its convenience and to its highly satisfactory performance as a disinfectant, which has been established by decades of use. It has been so successful that freedom from epidemics of waterborne diseases is now virtually taken for granted. As stated in Drinking Water and Health, "chlorination is the standard of disinfection against which others are compared."

However, the discovery that chlorination can result in the formation of trihalomethanes (THM's) and other halogenated hydrocarbons has prompted the reexamination of available disinfection methodology to determine alternative agents or procedures.

The method of choice for disinfecting water for human consumption depends on a variety of factors. These include:

- Its efficacy against waterborne pathogens (bacteria, viruses, protozoa, and helminths);

- The accuracy with which the process can be monitored and controlled;

- Its ability to produce a residual that provides an added measure of protection against possible post treatment contamination resulting from faults in the distribution system;

- The aesthetic quality of the treated water;

- The availability of the technology for the adoption of the method on the scale that is required for public water supplies.

Economic factors will also play a part in the final decision; however, this study is confined to a discussion of the five factors listed above as they apply to various disinfectants.

The propensity of various disinfection methods to produce by-products having effects on health (other than those relating to the control of infectious diseases) and the possibility of eliminating or avoiding these undesirable by-products are also important factors to be weighed when making the final decisions about overall suitability of methods to disinfect drinking water.

Organization of the Study

The general considerations noted in the immediately following material should be borne in mind when considering each method of disinfection. Available information on the obvious major candidates for drinking water disinfection—chlorine, ozone, chlorine dioxide, iodine, and bromine—is then evaluated for each method individually in the following sections. Other less obvious possibilities are also examined to see if they have been overlooked unjustly in previous studies or if it might be profitable to conduct further experimentation on them. Disinfection by chloramines is dealt with in parallel with that effected by chlorine because of the close relationship the former has to chlorine disinfection under conditions that might normally be encountered in drinking water treatment.

The evaluations in this report are not exhaustive literature reviews but, rather, are selections of the studies that, in the judgment of the committee, provide the most accurate and relevant information on the biocidal activities of each method of disinfection. The analytical methods that are described in this report are those that are most likely to be used by persons involved in disinfection research or water treatment. A review of all existing analytical methods, some of which may be more sophisticated than those described below, would be impractical within the constraints of time and space available and is not within the scope of this document.

After the methods of disinfection are examined individually, their major characteristics and biocidal efficacy are compared by means of summary tables and $c \cdot t$ (concentration, in milligrams per liter, times contact time, in minutes) values required for similar inactivations under identical conditions. The conclusions of the study are then recorded on the basis of this evidence.

General Aspects of Disinfection

In any comparison of disinfection methods, certain considerations should be discussed at the outset since they are relevant to most, if not all, methods. The quality of the raw water (i.e., its content of solids and material that will react with the disinfectant), treatment of the water prior to disinfection, and the manner in which the disinfectant is applied to the water will directly affect the efficacy of all disinfectants. Equally applicable to all methods are appropriate standards for verifying the adequacy of disinfection, differences in response to disinfectants between organisms that were obtained directly from the field and those that have been acclimated to laboratory culture, and the maintenance of portability from treatment plant to the consumer's tap. The use of chlorination as presented in examples in the following pages does not imply that it is necessarily the method of choice. Rather, this method has been studied more thoroughly than other methods.

Raw Water Quality

In addition to potential pathogens, raw water may contain contaminants that may interfere with the disinfection process or may be undesirable in the finished product. These contaminants include inorganic and organic molecules, particulates, and other organisms, e.g., invertebrates. Variations among these contaminants arise from differences in regional geochemistry and between ground- and surface-water sources.

Disinfectant Demand

Many inorganic and organic molecules that occur in raw water exert a "demand," i.e., a capacity to react with and consume the disinfectant. Therefore, higher "demand" waters require a greater dose to achieve a specific concentration of the active species of disinfectant. This demand must be satisfied to ensure adequate biocidal treatment.

Ferrous ions, nitrites, hydrogen sulfide, and various organic molecules exert a demand for oxidizing disinfectants such as chlorine. The bulk of the nonparticulate organic material in raw water occurs as naturally derived humic substances, i.e., humic, fulvic, and hymatomelanic acids, which contribute to color in water. The structure of these molecules is not yet fully understood. However, they are known to be polymeric and to contain aromatic rings and carboxyl, phenolic, alcoholic hydroxyl, and methoxyl functional groups. Humic substances, when reacting with and consuming applied chlorine, produce chloroform $(CHCl_3)$ and other THM's. Water, particularly surface waters, may also contain synthetic organic molecules whose demand for disinfectant will be determined by their structure. Ammonia and amines in raw water will react with chlorine to yield chloramines that do have some biocidal activity, unlike most products of these side reactions. If chlorination progresses to the breakpoint, i.e., to a free-chlorine residual, these chloramines will be oxidized causing more added chlorine to be consumed before a specific free-chlorine level is achieved.

The nature of the demand reactions varies with the composition of the water and the disinfectant. Removal of the demand substances leaves water with a lower requirement for a disinfectant to achieve an equivalent degree of protection against transmission of a waterborne disease.

Physical and Chemical Treatments

Various treatments applied to raw water to remedy undesirable characteristics, e.g., color, taste, odor, or turbidity, may affect the ultimate microbiological quality of the finished water. Microorganisms may be physically removed or the disinfectant demand of the water altered.

Presedimentation to remove suspended matter, coagulation with alum or other agents, and filtration reduce the organic material in the raw water and, thus, the disinfectant demand. Removal of ferrous iron similarly reduces the demand for oxidizing disinfectants as will aeration, which eliminates hydrogen sulfide. Prechlorination to free chlorine residual is practiced early in the treatment sequence as one method to alter taste- and odor-producing compounds, to suppress growth of organisms in the treatment plant, to remove iron and manganese, and to reduce the interference of organic compounds in the coagulation process.

The necessity for these treatments or others is determined by the characteristics of the raw water. The selection of one of the various methods to achieve a particular result will be based upon cost-effectiveness in the particular situation. When chlorination is used, the application or point of application in the treatment sequence of some of the above-mentioned procedures can affect the undesirable THM content of the finished water.

Reduction of precursors in raw water by coagulation and settling prior to chlorination reduces final THM production. The Louisville Water Company reduced THM concentrations leaving the plant by 40%-50% by shifting the point of chlorination from the pre-sedimentation basin to the coagulation basin. The available information on these variations is limited, and a universally applicable procedure cannot be recommended in view of the diverse treatments required for different raw waters.

Particulates and Aggregates

To inactivate organisms in water, the active chemical species must be able to reach the reactive site within the organism or on its surface. Inactivation will not result if this cannot occur. Microorganisms may acquire physical protection in water as a result of their being adsorbed to the enormous surfaces provided by clays, silt, and organic matter or to the surfaces of solids created during water treatment, e.g., aluminum or ferric hydrated oxides, calcium carbonate, and magnesium hydroxide. Viruses, bacteria, and protozoan cysts may be adsorbed to these surfaces. Such particles, with the adsorbed microorganisms, may aggregate to form clumps, affording additional protection. Organisms themselves may also aggregate or clump together so that organisms that are on the interior of the clump are shielded from the disinfectant and are not inactivated. Organisms may also be physically embedded within particles of fecal material, within larger organisms such as nematodes, or, in the case of viruses, within human body cells that have been discharged in fecal material.

To disinfect water adequately, the water must have been pretreated, when necessary, to reduce the concentration of solid materials to an acceptably low level. The primary drinking water turbidity standard of 1 nephelometric turbidity unit (NTU) is an attempt to assure that the concentration of particulates is compatible with current disinfection techniques. Where it is possible to obtain lower turbidities, this is desirable.

Disinfection studies in which the complications of adsorbed organisms, aggregation, or embedment were thought to occur were excluded from this study. The conclusions in this report should not be extrapolated to such situations as the disinfection of turbid or colored waters.

The Importance of Residuals

Water supplies are disinfected through the addition or dosage of a chemical or physical agent. With a chemical agent, such as a halogen, a given dosage should theoretically impart a predetermined concentration (residual) of the active agent in the water. From a practical point of view, most natural waters exert a "demand" for the disinfectant, so that the residual in the water is less than the calculated amount based on the dosage. The decrease in residual, which is caused by the demand, is rapid in most cases, but it may be prolonged until the residual eventually disappears. In addition, the chemical agent may decompose spontaneously, thereby yielding substances having little or no disinfection ability and exerting no measurable residual. For example, ozone not only reacts with substances in water that exert a demand, but it also decomposes rapidly. To achieve microbial inactivation with a chemical agent, a residual must be present for a specific time. Thus, the nature and level of the residual, together with time of exposure, are important in achieving disinfection or microbial inactivation. Because the nature of the dosage-residual relationship for natural waters has not been and possibly cannot be reliably defined, the efficacy of disinfection with a chemical agent must be based on a residual concentration/time of exposure relationship.

Residual measurements are important and useful in controlling the disinfection process. By knowing the residual-time relationship that is required to inactivate pathogenic or infectious agents, one can adjust the dosage of the disinfecting agent to achieve the residual that is required for effective disinfection with a given contact time. Thus, the effectiveness of the disinfection process can be controlled and/or judged by monitoring or measuring the residual.

Following disinfection of a water supply at a treatment plant, the water is distributed to the consumers. A persistent residual is important for continued protection of the water supply against subsequent contamination in the distribution system. Accidental or mechanical failures in the distribution system may result in the introduction of infectious agents into the water supply. In the presence of a residual, disinfection will continue and, as a result, offer continued protection to the users. Physical agents such as radiation may provide effective disinfection during application, but they do not impart any persistent residual to the water.

The dosage of a chemical agent that is used to effect microbial inactivation should not be so great that it imparts a health hazard to the water consumer. From another point of view, the aesthetic quality of the finished water should not be impaired by the dosage of the chemical agent or the residual that is required for effective disinfection. These qualities might include discoloration of water from potassium permanganate $(KMnO_4)$ or iodine or problems of taste and odor from excessive chlorine.

Application of the Disinfectant

Optimum inactivation occurs when the disinfectant is distributed uniformly throughout the water. To disperse the chemical disinfectant when it is added to the water, it must be mixed effectively to assure that all of the water, however small the volume, receives its proportionate share of the chemical. Additions of a disinfectant at points in a flowing water stream, e.g., from submerged pipes, is seldom adequate to assure uniform concentration. In such cases, mechanical mixing devices are needed to disperse the disinfectant throughout the water. Disinfection by radiation treatment also requires good mixing to bring all of the water within the effective radiation distance.

Microbiological Considerations

Comparison of the biocidal efficacy of disinfectants is complicated by the need to control many variables, a need not realized in some early studies. Halogens in particular are significantly affected by the composition of the test menstruum and its pH, temperature, and halogen demand. For very low concentrations of halogen to be present over a testing period, halogen demand must be carefully eliminated. Different disinfectants may have different biocidal potential. In earlier work, analytical difficulties may have precluded defining exactly the species present, but new techniques allow the species to be defined for most disinfectants. Information on the species of disinfectant actually in the test system should be included in future reports on disinfection studies.

Investigators studying efficacy have usually adopted one of two extremes. Some have conducted carefully designed laboratory experiments with controls for as many variables as possible. Certain of these investigators have reduced the temperature to slow the inactivation reactions. Although these experiments yield good basic information and can be used to determine which variables are important, they often have little quantitative relationship to field situations. The other extreme, a field study or reconstruction of field conditions, is difficult to control. Moreover, their results are often not repeatable.

In addition to the variables noted above, prereaction of chemicals in the test system, the culture history of the organism being used, and the "cleanup" procedures applied to it may also affect the observed results. Despite these problems, there have been some attempts to standardize efficacy testing.

Model Systems and Indicator Organisms

A major factor that influences the evaluation of the efficacy of a particular disinfectant is the test microorganism. There is a wide variation in susceptibility, not only among bacteria, viruses, and protozoa (cyst stage), but also among genera, species, and strains of the microorganism. It is impractical to obtain information on the inactivation by each disinfectant for each species and strain of pathogenic microorganism of importance in water. In addition, interpretation of the data would be confounded by the condition and source of the test microorganism (e.g., the degree of aggregation and whether the organisms were "naturally occurring" or laboratory preparations), the presence of solids and particulates, and the presence of materials that react with and consume the disinfectant.

The overwhelming majority of the literature on water disinfection concerns the inactivation of model microorganisms rather than the pathogens. These disinfectant model microorganisms have generally been nonpathogenic microorganisms that are as similar as possible to the pathogen and behave in a similar manner when exposed to the disinfectant. The disinfectant model systems are simpler, less fastidious, technically more workable systems that provide a way to obtain basic information concerning fundamental parameters and reactions. The information gained with the model systems can then be used to design key experiments in the more difficult systems. The disinfection model microorganism should be clearly distinguished from the indicator organism. The indicator microorganism, as defined in Drinking Water and Health, is a "microorganism whose presence is evidence that pollution (associated with fecal contamination from man or other warm-blooded animals) has occurred." Following are criteria for the indicator microorganism:

- The indicator should always be present when fecal material is present and absent in clean, uncontaminated water.

- The indicator should die away in the natural aquatic environment and respond to treatment processes in a manner that is similar to that of the pathogens of interest.

- The indicator should be more numerous than the pathogens.

- The indicator should be easy to isolate, identify, and enumerate.

Only a restrictive application of the second criterion is necessary for a disinfection model. The response of the test microorganism to the disinfectant must be similar to that of the pathogen that it is intended to simulate. The disinfection model is not meant to function as an indicator microorganism.

During the latter part of the nineteenth century, investigators recognized the presence of a group of bacteria that occured in large numbers in feces and wastewater. The most significant member of this group (currently called the coliform group) is Escherichia coli. Since the late nineteenth century, this coliform group has served as an indicator of the degree of fecal contamination of water, and E. coli has been used routinely as a disinfection model for enteric pathogens. Butterfield and co-workers provided information on the inactivation of E. coli and other enteric bacterial pathogens with chlorine and chloramines. At pH values above 8.5, all strains of E. coli were more resistant to free chlorine than were Salmonella typhi strains. At pH values of 6.5 and 7.0, strains of S. typhi were more resistant. Only slight differences between the two genera were found when chloramines were used as the disinfectant. The bactericidal activity of chloramine was noticably less than that of free chlorine.

Bacteria of the coliform group, especially E. coli, have proved useful as an indicator and disinfection model for enteric bacterial pathogens but are poor indicators and disinfection models for nonbacterial pathogens. E. coli has been observed to be markedly more susceptible to chlorine than certain enteric viruses and cysts of pathogenic protozoa.

The bacterial viruses of E. coli have received increased attention as possible disinfection models and indicators of enteric viruses in water and wastewater. At present, the data to justify the bacterial viruses as indicators for enteric viruses are limited and inconsistent. However, there is a growing body of knowledge on the utilization of bacterial viruses as disinfection models.

Hsu and Hsu et al. first reported the use of the f2 virus as a model for disinfection studies with iodine. They showed that inactivation of both the f2 virus and poliovirus 1 were inhibited by increasing concentrations of iodide ion and that both f2 RNA and poliovirus 1 RNA were resistant to iodination.

Dahling et al. compared the inactivation of two enteric viruses (poliovirus 1 and coxsackievirus A9), two DNA phages (T2 and T5), two RNA phages (f2 and MS2), and E. coli ATCC 11229 under demand-free conditions with free chlorine at pH 6.0. They found enteric viruses to be most resistant to free chlorine followed by RNA phages, E. coli, and the T phages.

Shah and McCamish compared the resistance of poliovirus 1 and the coliphages f2 and T2 to 4 mg/liter combined residual chlorine. The f2 virus was shown to be more resistant to this form of chlorine than poliovirus 1 and T2 coliphage.

Cramer et al. compared the inactivation of poliovirus 3 (Leon) and f2 with chlorine and iodine in buffered wastewater. Both viruses were treated together in the same reaction flask, thereby eliminating any inherent differences due to virus preparations and replicate systems. In wastewater effluent at pH 6.0 and 10.0 with a 30 mg/liter dosage of halogen under prereacted (halogen added to wastewater, allowed to react, viruses added at zero time) and dynamic (viruses added to wastewater, halogen added at zero time) conditions, f2 was, in each case, at least as or more resistant to chlorine and iodine than poliovirus 1. The f2 virus appears to be more sensitive to free chlorine but more resistant to combined chlorine than poliovirus 1 is.

Neefe et al. observed that the agent of infectious hepatitis was inactivated by breakpoint chlorination (free chlorine) but not completely inactivated by combined chlorine.

Engelbrecht et al. reported that the use of a yeast (Candida parapsilosis) and two acid-fast bacteria (Mycobacterium fortuitum and Mycobacterium phlei) may provide suitable disinfection models. They observed that the yeast was more resistant to free chlorine than were poliovirus 1 and the enteric bacteria under all conditions tested. The acid-fast bacilli were most resistant.

There is no generally accepted disinfection model for protozoan cysts. In disinfection studies for protozoan diseases, investigators have used the pathogen or its cysts. Work with such systems is, however, generally difficult.

The use of disinfection models provides useful information that is helpful to the comparison of the relative efficiencies of various disinfectants in the laboratory and in controlled field investigations. Strains of E. coli have been used extensively as models for enteric pathogenic bacteria. While not as widely accepted, the bacterial viruses of E. coli are used as disinfection models for enteric viruses. The difficulty of available methods has limited the number of disinfection studies with protozoan cysts.

Laboratory Cultures versus Naturally Occurring Organisms

The resistance or sensitivity to disinfectants of some bacteria (e.g., E. coli) in the laboratory may bear very little resemblance to their responses in nature. This is true in spite of the fact that standardized procedures govern the conditions under which cells are grown, harvested, washed, etc., when they are used as inocula. Examples of such differences range from Gram-negative bacteria and their comparative resistance to disinfectants in general to Gram-positive bacterial spores and heat resistance and to halogen resistance of Entamoeba histolytica cysts from simian hosts as opposed to those grown in in-vitro systems. Presumably, the mechanisms creating this phenomenon among these three groups vary widely.

The comparative resistance to disinfectants among Gram-negative bacteria varies greatly. A good example of this is the study of Favero and Drake. They first applied the term "naturally occurring" to certain Gram-negative bacteria with the potential for rapid growth in water. They observed that Pseudomonas alcaligenes, a common bacterial contaminant in iodinated swimming pools, could grow well in swimming pool waters that had been sterilized by membrane filters and rendered free of iodine or chlorine. Starting with contaminated swimming pool water that contained a variety of bacteria, they isolated a pure culture of P. alcaligenes by an extinction-dilution technique in which filter-sterilized swimming pool water was used as the diluent and growth medium. Since these cells had been isolated in pure culture without exposure to conventional laboratory culture media, they were referred to as "naturally occurring" P. alcaligenes. Subsequent tests showed that these naturally occurring cells were significantly more resistant to free iodine than were cells of the same organism that had been subcultured one time on trypticase soy agar. In fact, standard disinfectant tests using the cells that had been subcultured on an enriched laboratory medium suggested that P. alcaligenes should never be found in pools that had been disinfected even minimally with iodine. This was obviously an erroneous assumption. The discovery that naturally occurring cells were extremely resistant to iodine explained the relatively high concentrations of P. alcaligenes that accumulated in pool water that had been iodinated for several weeks.

Subsequently, Favero et al. and Carson et al. published a series of papers showing that Pseudomonas aeruginosa could grow rapidly in distilled water, which they obtained from hospitals, and could reach high concentrations of cells that remained stable for a long time. Naturally occurring cells that were grown in distilled water reacted quite differently to chemical and physical stresses than did cells grown on standard laboratory culture media. For example, naturally occurring cells of P. aeruginosa were significantly more resistant to chlorine, quaternary ammonium compounds, and alkaline glutaraldehyde than were subcultured cells.

In halogen-disinfected waters, naturally occurring bacteria can be from one to two orders of magnitude more resistant to the disinfectant than cells of the same organism that had been subcultured on conventional laboratory culture media. Since standard disinfectant testing necessarily employs subcultured and washed bacterial cells, a false sense of confidence may be created if these data are used as an absolute criterion for the dilution of a disinfectant. These results could explain the frequent discrepancies between tests that are performed under laboratory conditions and those that are performed under field conditions.

If bacteria could be used in their naturally occurring state, one might explore the possibility of bridging the gaps between laboratory and field conditions by using this experimental system. The ability of some Gram-negative bacteria to grow in water makes it possible to produce and control large numbers of cells for such studies.

More difficult to answer is the more basic question of why naturally occurring cells of Gram-negative water bacteria become more sensitive to disinfectants when grown in a rich medium than the same strain when grown in water. One would expect the reverse to occur. Milbauer and Grossowicz showed that cells of E. coli were much more sensitive to chlorine when grown on a medium of glucose mineral salts than when grown on nutrient agar. Since Favero and Drake reported that filter-sterilized dehalogenated swimming pool water could be considered a minimal medium, one would expect that P. alcaligenes cells that were grown in this environment would be less resistant to iodine than those grown in trypticase soy broth. This phenomenon has not been explained. Evidently it is not primarily a genetic response since the extreme difference in iodine resistance occurs with one subpassage on trypticase soy agar.

Over the years various investigators have tried without success to "train" bacteria to become more resistant to chlorine and iodine. This failure is not surprising, because, if halogens are truly a general cytoplasmic poison that affects primarily the sulfhydryl groups of enzymes, it would be very difficult for an organism to modify its physiology to the extent that it becomes resistant, very unlike the situation with antibiotics and bacteria. Consequently, the extreme resistance or differing resistances of naturally occurring bacteria can be attributed only to "environmental" factors and, perhaps, to the different compositions of cell walls and membranes. However, there have been no data to substantiate this hypothesis.

Despite the questions that have been raised by differences in the behavior of organisms under both laboratory and field conditions, valuable comparative information can be obtained from studies of disinfectants that are conducted in similar laboratory systems.

Chlorine and Chloramines

Chlorine is a strong oxidizing disinfectant that has been used to treat drinking water supplies for more than 60 yr. The gas was named "chlorine" after the Greek word for green, "chloros," because of its characteristic color. About 1800, chlorine gas was used as a general disinfectant in both France and England. In the United States, electrolytically produced chlorine was first used directly for water disinfection for only a week or two in 1896 at the Louisville Experimental Station in Kentucky. The first continuous municipal application of chlorine (as sodium hypochlorite [NaOCl]) to water in the United States occurred in 1908 at Jersey City, New Jersey. This was followed in 1912 by the first full-scale use of liquid chlorine for water disinfection at Niagara Falls, New York, where solution-feed equipment was used. This use of chlorine successfully eliminated recurring outbreaks of typhoid fever. In 1913, improved solution-feed equipment was developed to measure chlorine gas, dissolve it in water, and apply the solution to the water supply. This equipment was first installed at Boonton, New Jersey, where it replaced the use of sodium hypochlorite. Other equally significant historical occurrences led to the eventual addition of chlorine to drinking water in most of the United States for disinfection to destroy or inactivate pathogenic microorganisms.

Chemistry of Chlorine in Water

Chlorine has an atomic number of 17, a melting point of -102 °C, a boiling point of -35 °C, and an oxidation potential of -1.36 V at 25 °C $\left(i.e., Cl^- \rightleftharpoons Cl_2 + 2e^- \right)$. It is a green-yellow gas at room temperature.

When chlorine is added to water, the following chemical reactions occur:

$$Cl_2 + H_2O \rightleftharpoons HOCl + H^+ - Cl^-$$

$$HOCl \rightleftharpoons H^+ + OCl^-$$

Extremely little molecular chlorine $\left(Cl_2 \right)$ is present at pH values greater than pH 3.0 and total chlorine concentrations of less than ~1,000 mg/liter. The hypochlorous acid (HOCl) that is produced further ionizes to form hypochlorite ion $\left(OCl^- \right)$ and hydrogen ion $\left(H^+ \right)$ (Reaction $HOCl \rightleftharpoons H^+ + OCl^-$). The dissociation of hypochlorous acid is dependent chiefly upon pH and, to a much lesser extent, temperature, with almost 100% hypochlorous acid present at pH 5 and almost 100% hypochlorite ion present at pH 10. Free available chlorine refers to the concentration of hypochlorous acid and hypochlorite ion, as well as any molecular chlorine existing in a chlorinated water.

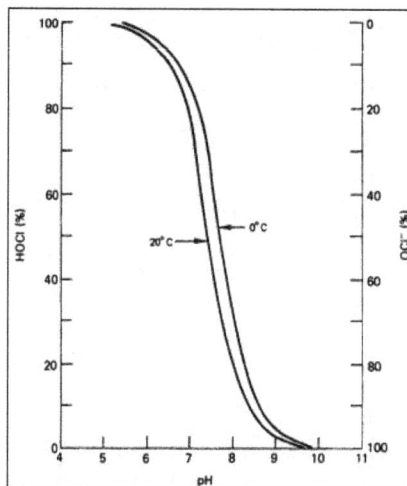

Effect of pH on quantities of hypochlorous acid (HOCl) and hypochlorite ion $\left(OCl^- \right)$ that are present in water.

Chloramine Formation: Inorganic Chloramines

During chlorination of a water supply for disinfection, chlorine will react with any ammonia $\left(NH_3 \right)$ in the water to form inorganic chloramines.

Furthermore, ammonia is sometimes deliberately added to chlorinated public water supplies to provide a combined available chlorine residual, i.e., inorganic chloramines. Chlorine will also react with organic amines. The organic chloramines that are produced are considered encompassed in the term "combined available chlorine."

Although inorganic chloramines are less effective oxidizing and disinfecting agents than hypochlorous acid and hypochlorite ion, they are more stable. Consequently, they will produce a residual in water that will persist for a longer time.

Inorganic chloramines are formed when hypochlorous acid reacts with ammonia:

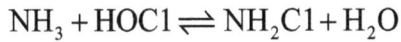

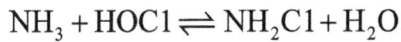

$$NH_3 + HOCl \rightleftharpoons NH_2Cl + H_2O$$

$$NH_3 + HOCl \rightleftharpoons NH_2Cl + H_2O$$

The chloramine that is formed in the reaction depends upon the ratio of ammonia to hypochlorous acid and the pH of the system. Dichloramine $(NHCl_2)$ is the predominant form of chloramine at a 1:1 molar ratio of ammonia to chlorine at pH values of 5 and below, whereas at pH values of 9 and above, monochloramine (NH_2Cl) predominates. Figure shows the proportions of monochloramine and dichloramine formed for pH values of 4 to 9 and temperatures of 0 °C, 10 °C, and 25 °C.

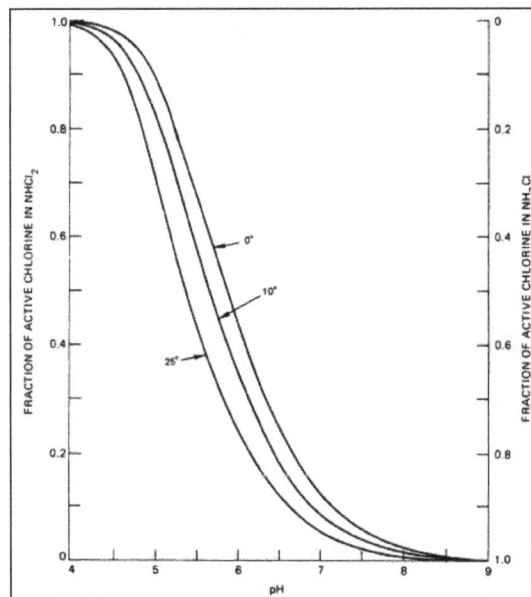

Proportions of mono- and dichloramine (NH_2Cl and $NHCl_2$) in water chlorination with equimolar concentrations of chlorine and ammonia.

Organic Chloramines

Chlorine is also known to combine slowly with organic or albumenoid nitrogen (amines) to form organic chloramines:

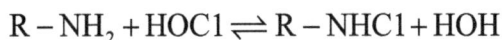

$$R - NH_2 + HOCl \rightleftharpoons R - NHCl + HOH$$

Although the reaction between organic amines and chlorine is generally considered to be slow, organic chloramines may be formed, thereby producing a stable combined available chlorine residual after many hours of contact. It is generally accepted that most organic chloramines have little disinfecting capability, i.e., less than the inorganic chloramines.

Breakpoint Chlorination

Hypochlorous acid and other chlorine compounds having disinfecting ability by virtue of their being oxidizing agents will oxidize sulfites (SO_3^{2-}), sulfides (S^-), and ferrous (Fe^{2+}) or manganous (Mn^{2+}) ions. The disinfecting species are reduced, and the products have no disinfecting activity. All of the interfering compounds that destroy the disinfecting ability of the added chlorine exert a "chlorine demand," which may be defined as the difference between the amount of chlorine applied and the quantity of free or combined available chlorine residual measured in the water at the end of a specified contact period. When chlorine is added to water with no chlorine demand, a linear relationship is established between the chlorine dosage and the free chlorine residual.

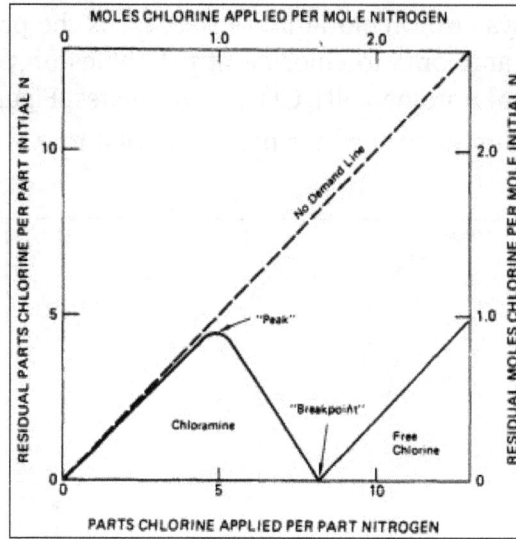

Diagrammatic representation of completed breakpoint reaction.

However, when increasing amounts of chlorine are added to water containing reducing agents and ammonia, the so-called breakpoint phenomenon occurs. The breakpoint is that dosage of chlorine that produces the first detectable amount of free available chlorine residual.

When chlorine is added to water, it reacts with any reducing agents and ammonia that are present. It is believed that chlorine reacts first with the reducing agents. Since the chlorine is destroyed, no measurable residual is produced. Following the oxidation of these reducing agents, e.g., sulfides, sulfites, nitrites (NO_2^-), and ferrous ions, the chlorine will react with ammonia to form inorganic chloramines. The quantity of monochloramine and dichloramine that is formed is determined primarily by the pH of the water and the ratio of chlorine to ammonia. When the ratio by weight is less than 5:1, or the molar ratio is less than 1:1, and the pH is in the range of 6.5 to 8.5, the combined available chlorine residual is probably due primarily to monochloramine (Reaction $NH_3 + HOCl \rightleftharpoons NH_2Cl + H_2O$). With additional chlorine, the ratio of chlorine to ammonia changes with the result that the monochloramines are converted to dichloramines (Reaction $NH_3 + HOCl \rightleftharpoons NH_2Cl + H_2O$). When all of the ammonia has been reacted, a free available chlorine residual begins to develop. As the concentration increases, the previously formed chloramines are oxidized to nitrous oxide (N_2O), nitrogen trichloride (NCl_3), and nitrogen (N_2). The reactions leading to the formation of these oxidized forms of nitrogen destroy the combined available chlorine residual so that the measurable residual in the water actually decreases. Upon completion of the oxidation of all the chloramines, the addition of more chlorine creates the breakpoint phenomenon. At the breakpoint dosage, some resistant chloramines may still be present, but at such small concentrations that they are unimportant.

As pointed out by Morris, the occurrence of reactions giving rise to the "breakpoint" is most rapid in the pH range 7.0 to 7.5. At greater and lesser pH values, it becomes slower and less distinct, e.g., at pH's < 6 or > 9 the concept of "breakpoint" is not significant. In the pH range 7.0 to 7.5 the "breakpoint" is about half developed within 10 min at 15 °C to 20 °C and is then substantially completed within about 2 hr.

Analytical Methods and their Evaluation

Standard Methods lists six acceptable methods for the determination of chlorine residuals in natural and treated waters: iodometric methods, amperometric titration, the stabilized neutral orthotolidine (SNORT) method, the ferrous diethyl-p-phenylenediamine (DPD) method, the DPD colorimetric method, and the leuco crystal violet (LCV) method.

The amperometric, LCV, DPD, and SNORT methods are unaffected by dichloramine concentrations in the range of 0 to 9 mg/liter (as Cl_2) in the determination of free chlorine. If nitrogen trichloride is present, it reacts partially as free available chlorine in the amperometric, DPD, and SNORT methods. Nitrogen trichloride does not interfere with the LCV procedure for free chlorine. The sample color and turbidity may interfere with all colorimetric procedures. Thus, a compensation must be made. Also, organic contaminants in the sample may produce a false-free chlorine reading in most colorimetric methods.

Standard Methods contains data on the precision and accuracy of the methods used in the measurement of chlorine. These data were obtained from participating laboratories by the Analytical Reference Service, which then operated in an agency that preceded the Environmental Protection Agency. However, as noted in Standard Methods, these results are valuable only for comparison of the methods tested, and many factors, such as analytical skill, recognition of known interferences, and inherent limitations, determine the reliability of any given method. Moreover, some oxidizing agents, including free halogens other than chlorine, will appear quantitatively as free chlorine. This is also true of chlorine dioxide. Also, some nitrogen trichloride may be measured as free chlorine. The actions of interfering substances should be familiar to the analyst because they affect a particular method.

Although orthotolidine (i.e., orthotolidine and orthotolidine arsenite) methods have been widely used in many disinfection studies, they are omitted from the 14th edition of Standard Methods primarily because of their inaccuracy and high overall (average) total error in comparison with other available methods.

Research studies on disinfection are restricted by the limitations that are inherent in the methods themselves or by poor selection of methods by the investigator. The chemical conditions of the test water have not always been well-defined. The types of titratable chlorine, i.e., free (hypochlorous acid or hypochlorite ion) or combined (mono- or dichloramine) in the chlorinated water, have not always been differentiated, and the rates of microbial destruction or inactivation have not always been studied in experimental systems with little or no chlorine demand. In fact, reports prior to the 1940's have been especially difficult to interpret, because reliable test methods for distinguishing between free and combined chlorine, between hypochlorous acid and hypochlorite ion, and between mono- and dichloramine in solution were not developed until the 1950's. For example, many earlier researchers claimed to have tested mono- and dichloramine by controlling the pH and the ratio of chlorine to nitrogen. They used methods such as the orthotolidine or thiosulfate titrations to determine total chlorine residual. Much of this early work is now questionable, since it was not possible to detect free chlorine contamination in their chloramine solutions or the quantitative ratios between the mono- and dichloramine tested. in addition, these earlier studies had high chlorine demand in the test systems.

Some more contemporary studies have lacked quantitated information on chlorine residual and/or types of chlorine present in the test systems.

Biocidal Activity

In the absence of reducing agents, inorganic ammonia, and organic amines, the addition of chlorine to municipal water supplies will result in free available residual chlorine, represented by the hypochlorous acid or hypochlorite ion. The pH determines the relative amounts of each species. However, inorganic chloramines will be formed if the background level of ammonia in the water supply is significant or if ammonia is intentionally added during treatment. If such is the case, monochloramine would predominate due to the alkaline pH of most finished water.

In 1966, Feng proposed that the active forms of chlorine would exhibit disinfection properties in the following descending order:

$$Cl_2 > HOCl > OCl^- > NHCl_2 > NH_2Cl > R - NHCl$$

Butterfield et al. published the first treatise on the use of chlorine demand-free water for studies of water disinfection. They proposed that to study the disinfectant capacity of any chlorine species, the test medium must meet certain exacting criteria. It must be nontoxic to bacteria except for the variables under study such as chlorine and pH, well buffered at the desired pH, free of all ammonia and organic matter capable of forming chlorine-addition products, free of background chlorine, and of such a nature that calculated additions of chlorine are recoverable after 5 min without a loss in residual and that free chlorine must still be present several hours after contact.

Most studies of combined chlorine have dealt with poorly defined mixtures of mono- and dichloramine. Also, test conditions have often been inadequately defined, poorly controlled, or both.

Efficacy against Bacteria

Free Chlorine (HOCl and OCl⁻)

Butterfield et al. studied percentages of inactivation as functions of time for E. coli, Enterobacter aerogenes, Pseudomonas aeruginosa, Salmonella typhi, and Shigella dysenteriae. They used different levels of free chlorine at pH values ranging from 7.0 to 10.7 and two temperature ranges—2 °C to 5 °C and 20 °C to 25 °C. Their work is of great importance, since very few other studies have been conducted that dealt with the action of disinfectants on pathogens. Generally, they found that the primary factors governing the bactericidal efficacy of free available chlorine and combined available chlorine were:

- The time of contact between the bacteria and the bactericidal agent, i.e., the longer the time, the more effective the chlorine disinfection process;

- The temperature of the water in which contact is made, i.e., the lower the temperature, the less effective the chlorine disinfecting activity;

- The pH of the water in which contact is made, i.e., the higher the pH, the less effective chlorination.

Thus, the test bacteria will be killed more rapidly at lower pH values and at higher temperatures. Since hypochlorous acid would predominate at lower pH's, the data of Butterfield et al. show that it is a better bactericide than the hypochlorite ion. For example, to produce a 100% inactivation of an initial inoculum of 8×10^5 E . coli in 400 ml of sterile chlorine demand-free water (2,000/ml) at 20 °C-25 °C with a chlorine level of 0.046 to 0.055 mg/liter, 1.0 min was required at pH 7.0, but at pH 8.5, 9.8, and 10.7, between 20 and 60 min of exposure were needed. At higher concentrations of chlorine, i.e., from 0.1 to 0.29 mg/liter, exposure of 1.0 min was required at pH 7.0, 10 min at pH 8.5, 20 min at 9.8, and 60 min at 10.7. A similar pH effect was noted for S. typhi.

Unfortunately, Butterfield et al. lifted their cells from agar slants but failed to wash them in demand-free water. The cells probably carried trace amounts of albumenoid nitrogen from the slants to the test flasks, thereby creating the small chlorine demand that the investigators had tried so carefully to avoid. The effect of such a trace amount of chlorine demand would be most apparent in test solutions with very low chlorine levels. In studies using approximately 0.1 mg/liter or less free chlorine at decreasing pH values, Butterfield et al. observed that the disinfection of the organisms required a very long time. This might indicate interference at the low levels due to the formation of combined chlorine.

Under very exact controlled test conditions of pH and temperature and using chlorine demand-free buffer systems, Scarpino et al. observed that at 5 °C E. coli was 99% destroyed by 1.0 mg/liter hypochlorous acid at pH 6 in less than 10 s and at pH 10 by 1.0 mg/liter hypochlorite ion in about 50s. Their studies, which totally eliminated any form of combined chlorine from the test solutions, indicated that hypochlorous acid was approximately 50 times more effective than the hypochlorite ion as a bactericide. Fair et al. and Berg analyzed the data of Butterfield et al. on the destruction of E. coli. They reported that hypochlorous acid was 70-80 times as bactericidal as hypochlorite ion.

Engelbrecht et al. investigated new microbial indicators of disinfection efficiency. In their chlorination studies, they noted the following decreasing order of resistance to free chlorine at pH values of 6, 7, and 10 and at 5 °C and 20 °C: acid-fast bacteria > yeasts > poliovirus > Salmonella typhimurium > E. coli.

Monochloramine NH_2Cl

In 1948, Butterfield summarized previous results on the bactericidal properties of chloramines (and free chlorine) in water at pH values ranging from 6.5 to 10.7 and in two temperature ranges—2 °C to 5 °C and 20 °C to 25 °C. The test bacteria included strains of Escherichia coli, Enterobacter aerogenes, Pseudomonas aeruginosa, Salmonella typhi, and Shigella dysenteriae. Although he admitted that adequate tests for separate determination of free and combined chlorine forms were not used in these studies, the solutions were vigorously prepared to ensure the exclusions of free chlorine. Chloramines were determined using orthotolidine; readings made after 10 to 30 s at 20 °C gave free chlorine levels, and those after standing for 10 min at 20 °C were recorded as total residual chlorine. Since no free chlorine was reported (and should not have been found, according to the authors), the 10-min readings of total residual chlorine were also those of total chloramine levels. No distinction could be made between monochloramine and dichloramine. However, he estimated that at pH 6.5, 7.0, 7.8, 8.5, 9.5, and 10.5 the chloramines were present as monochloramine at 35%, 51%, 84%, 98%, 100%, and 100%, respectively. The balance was believed to be dichloramine.

Butterfield and his associates found that chloramine disinfection was always slower than that for free chlorine. For example, in order to achieve a 100% inactivation of the initial number of bacteria tested after 60 min of contact time at 20 °C, 0.6 mg/liter chloramine was required at pH 7.0 and 1.2 mg/liter at pH 8.5. At 4 °C, a 100% inactivation required 1.5 mg/liter chloramine at pH 7.0 and 1.8 mg/liter at pH 8.5. However, only 0.03 to 0.06 mg/liter free chlorine was needed at pH ranges of 7.0 and 8.5 at either 4 °C or 22 °C to achieve 100% inactivation in 20 min.

The bactericidal effects of monochloramine alone were confirmed by Siders et al. in 1973. They found that at 15 °C E. coli was 99% destroyed in approximately 20 min using 1.0 ppm monochloramine in pH 9 borate buffer. Since E. coli was less resistant to monochloramine than were the animal viruses tested, Siders et al. questioned the validity of using E. coli as an indicator organism for measuring the viral quality of a chlorinated water supply. Chang had previously calculated that 4 mg/liter monochloramine would be needed to give 99.999% reduction of E. coli bacterium in 10 min at 25 °C.

Dichloramine NHCl$_2$

Chang calculated that 1.2 mg/liter dichloramine would be needed to give 99.999% reduction of enteric bacteria in 10 min at 25 °C.

In carefully conceived studies, Esposito et al. examined the destruction rates of test organisms in contact with dichloramine in demand-free phthalate buffer at pH 4.5 and 15 °C. Figure shows the comparisons that were made among enteroviruses (poliovirus 1 and coxsackievirus A9), the bacteriophage $\Phi X - 174$, and E. coli (ATCC 11229).

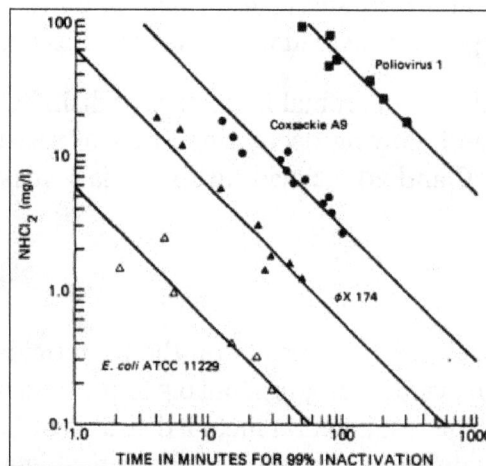

Inactivation of various microorganisms with dichloramine (NHCl$_2$) at pH 4.5 and 15 °C.

From a review of the literature and an analysis of the data, Chang calculated that the relative bactericidal efficiency of dichloramine to monochloramine was 3.3:1. However, Esposito and Esposito et al. showed experimentally that Chang's estimate was conservative. They found that dichloramine was 35 times more bactericidal than was monochloramine, not 3.3. At 15 °C, poliovirus 1 was 17 times more resistant than coxsackievirus A9, 83 times more resistant than $\Phi X - 174$, and 1,700 times more resistant than E. coli to dichloramine. They observed that dichloramine was a better bactericide than monochloramine.

Organic Chloramines

These chlorine derivatives exhibit some bactericidal activity, but markedly less than either free chlorine or the inorganic chloramines.

The bactericidal efficiency of hypochlorous acid, the hypochlorite ion, monochloramine, and dichloramine have been accurately defined in recent years by investigators using rigidly controlled test conditions. The order of disinfection efficiency presented by Feng has been confirmed. Comparative c · t values are shown in table.

Table: Dosages of Various Chlorine Species Required for 99% Inactivation of Escherichia Coli and Poliovirus 1.

Test Microorganism	Disinfecting Agent	Concentration, mg/liter	Contact Time, min	c·t	pH	Temperature, °C
E. coli	Hypochlorous acid (HOCl)	0.1	0.4	0.04	6.0	5
	Hypochlorite ion (OCl⁻)	1.0	0.92	0.92	10.0	5
	Monochloramine (NH$_2$Cl)	1.0	175.0	175.0	9.0	5
		1.0	64	64.0	9.0	15
		1.2	33.5	40.2	9.0	25
	Dichloramine (NHCl$_2$)	1.0	5.5	5.5	4.5	15
Poliovirus 1	Hypochlorous acid (HOCl)	1.0	1.0	1.0	6.0	0
		0.5	2.1	1.05	6.0	5
		1.0	2.1	2.1	6.0	5
		1.0	1.0	1.0	6.0	15
	Hypochlorite ion (OCl⁻)	0.5	21	10.5	10.0	5
		1.0	3.5	3.5	10.0	15
	Monochloramine (NH$_2$Cl)	10	90	900	9.0	15
		10	32	320	9.0	25
	Dichloramine (NHCl$_2$)	100	140	14,000	4.5	5
		100	50	5,000	4.5	15

Efficacy against Viruses

In reviewing disinfection of enteroviruses in water, Clarke and Chang excluded all studies on the inactivation of viruses by chlorine that was conducted before 1946. Their justification for this exclusion was the failure of these studies to differentiate between free and combined chlorine. Furthermore, they attributed the irregular virucidal results of some studies to the use of animal inoculation methods for assaying virus concentrations. For these same reasons, those studies have been omitted in this report. The advent of viral propogation techniques using tissue cultures enabled research of a more exacting nature to be performed, resulting in more precise virus inactivation data.

Free Chlorine (HOCl and OCl⁻)

Generally, enteroviruses are more resistant to free chlorine than are the enteric bacteria. For example, in what was probably the first well-defined study, Clarke and Kabler used purified coxsackievirus A2 to investigate viral inactivation in water by free chlorine. They carefully controlled their free chlorine residuals with a modified form of the orthotolidine test to determine total chlorine and an orthotolidine-arsenite method for free chlorine. (Combined chlorine was then calculated as the difference between "total" and "free" chlorine readings.) They measured virus recoveries by using suckling mice and the LD_{50} quantitation procedure. Their results indicated that inactivation times for the virus increased with increasing pH (6.9 to 9.0), decreasing temperatures (27 °C-29 °C to 3 °C-6 °C), and decreasing total chlorine concentration. They estimated that approximately 7 to 46 times as much free chlorine was required to obtain comparable inactivation of coxsackievirus A2 as was required for a suspension of E. coli cells. For instance, Butterfield et al. found that at pH 7.0 and at 2 °C to 5 °C, 99.9% of E. coli cells were inactivated in 5 min with 0.03 mg/liter of free chlorine. At approximately the same pH and temperature ranges, Clarke and Kabler observed 99.6% inactivation of coxsackievirus A2 in 5 min with 1.4 mg/liter of free chlorine, i.e., 46 times as much free chlorine as that required to inactivate E. coli cells. At a pH of 8.5 at 25 °C, 99.9% of E. coli cells were inactivated in 3 min with 0.14 mg/liter of free chlorine, while at a pH of 9.0 at 27 °C to 29 °C, 99.6% of the virus was inactivated in 3 min by 1.0 mg/liter of free chlorine. Thus, Clarke and Kabler›s work showed that 7 times as much free chlorine was required to inactivate the test coxsackievirus compared to the time necessary to kill the bacterium E. coli. In a subsequent study, Clarke et al. found that adenovirus type 3, E. coli, and Salmonella typhi were all inactivated or destroyed at approximately the same concentration of free chlorine.

In 1958, Weidenkopf reported his studies on the rate of inactivation of poliovirus 1 as a function of free available chlorine $\left(HOCl \text{ and } OCl^{-} \right)$ and pH at 0 °C. His results showed that 99% inactivation of poliovirus 1 was obtained with 0.10 mg/liter free chlorine in 10 min at pH 6.0 and of 0°C. At pH 6, most of the free chlorine should have been present as hypochlorous acid. Increasing the pH to 7.0 increased the time required for the same degree of inactivation by approximately 50%. At that pH, the free chlorine should have been a mixture containing predominantly hypochlorous acid and significant levels of hypochlorite ion. Both Weidenkopf and Clarke et al. indicated that an increase in pH (from 7.0 to 8.5, Weidenkopf; from 8.8 to 9.0, Clarke et al.) increased the inactivation time about sixfold. At these pH's, the free chlorine should have been present as mixtures of both hypochlorous acid and hypochlorite ion, but predominantly as the ion.

After comparing these studies, Clarke et al. concluded that at 0 °C to 6 °C poliovirus 1 and coxsackievirus A2 were considerably more resistant to hypochlorous acid than was E. coli, while adenovirus type 3 was more sensitive. For 99% destruction of E. coli, a 99-s contact time was required when the system was dosed with 0.1 mg/liter free chlorine as hypochlorous acid. The same percentage of the adenovirus was inactivated in approximately one-third of that time by the same concentration of hypochlorous acid. Under the same conditions, 8.5 min was required to inactivate 99% of the poliovirus, i.e., approximately 5 times the contact time required for E. coli. Coxsackievirus required a contact time for 99% inactivation in excess of 40 min, more than 24 times that required for E. coli.

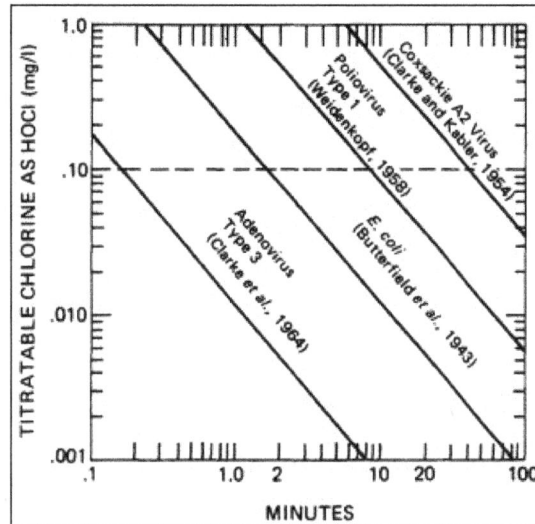

Concentration-time relationship for 99% destruction of E. coli and several viruses by hypochlorous acid at 0°-6 °C.

Clarke and Kabler and Clarke et al. reported that the time required for inactivation of coxsackievirus A2 and adenovirus increased with increasing pH, decreasing total chlorine concentrations, and decreasing temperatures. Clarke and Chang concluded from data in the literature that a 10 °C increase in temperature increased the rate of virus inactivation by a factor of 2 to 3.

Kelly and Sanderson reported that each of six enteric viruses possessed a different sensitivity to chlorine. Their results suggested that the inactivation of enteric viruses in water at pH 7.0 and 25 °C required a minimum free residual chlorine concentration of 0.3 mg/liter with a contact time of at least 30 min. With combined chlorine in water, a concentration of at least 9.0 mg/liter was necessary for a 99.7% inactivation of poliovirus at 25 °C and a pH of 7.0. Poliovirus 1 (strain MK 500) was the most resistant strain tested and coxsackievirus B5 the most sensitive. Poliovirus 1 (Mahoney strain), poliovirus 2, coxsackievirus B1, and poliovirus 3 were intermediate in resistance. The virucidal efficiency of hypochlorous acid was more than 50 times greater than that of the chloramines.

Liu et al. studied the manner in which 20 strains of human enteric viruses responded to free chlorine. They used Potomac River water that had been partially treated by coagulation with alum and filtration through sand. Chlorine was added to the water at one dosage, 0.5 mg/liter. The final pH was 7.8. They stored the sample at 2 °C. There was a wide range of resistance to chlorine by the viruses. The most sensitive virus was reovirus type 1, which required 2.7 min for inactivating 4 logs (99.99%) of the virus with 0.5 mg/liter of free chlorine. The most resistant, as judged by extrapolating the experimental data, was poliovirus 2, which required 40 min for the same degree of inactivation. Using actual experimental data, the most resistant virus was echovirus 12, which required a contact time of greater than 60 min for 99.99% inactivation. Liu et al. concluded from their extrapolated values that the reoviruses were the least resistant to chlorine treatment, that both adenoviruses and echoviruses were less resistant, and that the polioviruses and coxsackieviruses were the most resistant. However, assuming a 20-min contact time, most of the viruses tested at pH 7.8 and 2 °C would have been 99.99% inactivated with a free chlorine residual of 0.5 mg/liter.

Using six of the same virus strains studied by Liu et al., Engelbrecht et al. investigated the effect of pH on the kinetics of chlorine inactivation at 5.0 ± 0.2 °C. The suspending medium was buffered, chlorine demand-free, distilled-deionized water. Each virus stock was also prepared so as to be chlorine demand-free. Table summarizes the results, giving the chlorine levels used and the time required for two logs (99%) inactivation of the viruses at pH 6.0 and 7.8 in phosphate buffer and at pH 10.0 in borate buffer. Because of the use of two different buffer systems, i.e., at pH 6.0 and 10.0, virus inactivation was determined at pH 7.8 with each of the two buffer systems. The kinetics of inactivation of poliovirus 1 at pH 7.8, using both the phosphate and borate buffer, were the same. The results shown in table indicate that there is a significant difference in the time required for two logs inactivation for the various viruses at pH 6.0 and 10.0. In every case, the rate of inactivation at pH 10.0 was significantly less than at pH 6.0. The rank ordering in Table shows that there is also a wide range of sensitivity of related viruses to chlorine disinfection. For example, at pH 10.0, coxsackie B5 was 40 times more resistant than coxsackie A9. There are several cases in which the relative sensitivity to chlorine was altered (rank ordering) between pH 6.0 and 10.0, suggesting important effects of pH on the virion as well as on the chlorine species, i.e., hypochlorous acid versus. hypochlorite ion. This observation can be seen more clearly in Table in which the time required for two logs of inactivation of the various viruses and the ratio of inactivation times at pH 6.0 and 10.0 are compared. Even at pH 7.8, differences in relative sensitivity appear when ranked and compared to results at pH 6.0 or 10.0.

Table: Time Required for 99% Inactivation by Free Residual Chlorine at 5.0 °C ± 0.2 °C.

pH	Concentration of Free Chlorine, mg/liter	Virus Strain	Minutes for 99% Inactivation	Rank Ordering
6.00	0.46-0.49	Coxsackie A9 (Griggs)	0.3	1
6.00	0.48-0.49	Echo 1 (Farouk)	0.5	2
6.00-6.02	0.48-0.51	Polio 2 (Lansing)	1.2	3
6.00-6.03	0.38-0.49	Echo 5 (Noyce)	1.3	4
6.00	0.47-0.49	Polio 1 (Mahoney)	2.1	5
6.00-6.06	0.51-0.52	Coxsackie B5 (Faulkner)	3.4	6
		Coxsackie A9 (Griggs)	ND	
7.81-7.82	0.47-0.49	Echo 1 (Farouk)	1.2	1
		Polio 2 (Lansing)	ND	
7.79-7.83	0.48-0.52	Echo 5 (Noyce)	1.8	3
7.80-7.84	0.46-0.51	Polio 1 (Mahoney)	1.3	2
7.81-7.82	0.48-0.50	Coxsackie B5 (Faulkner)	4.5	4
10.00-10.01	0.48-0.50	Coxsackie A9 (Griggs)	1.5	1
10.00-10.40	0.49-0.51	Echo 1 (Farouk)	96.0	6
9.89-10.03	0.48-0.50	Polio 2 (Lansing)	64.0	4
9.97-10.02	0.49-0.51	Echo 5 (Noyce)	27.0	3
9.99-10.40	0.50-0.52	Polio 1 (Mahoney)	21.0	2
9.93-10.05	0.50-0.51	Coxsackie B5 (Faulkner)	66.0	5

Table: Comparison of Virus Inactivation by Free Residual Chlorine at pH 6.0 and 10.0, at 5.0°C ± 0.2 °C.

	Minutes for 99% Inactivation		
Virus Strain	pH 6.0	pH 10.0	Ratio
Coxsackie A9 (Griggs)	0.3	1.5	5
Echo 1 (Farouk)	0.5	96.0	192
Polio 2 (Lansing)	1.2	64.0	53
Echo 5 (Noyce)	1.3	27.0	21
Polio 1 (Mahoney)	2.1	21.0	10
Coxsackie B5 (Faulkner)	3.4	66.0	19

Combined Chlorine (NH$_2$Cl and NHCl$_2$)

Viral inactivation rates with chloramines have been found to be much slower than with free chlorine. For example, Kelly and Sanderson studied the effects of chlorine on several enteric viruses. They reported that at pH 7 at 25 °C-28 °C, 0.2-0.3 mg/liter free chlorine inactivated 99.9% of all test viruses in 8 min. At the same temperature and pH, combined chlorine at 0.7 mg/liter and at least 4 hr of contact time were needed to achieve 99.7% inactivation of the test viruses.

Although most viral inactivation studies with chloramines have not differentiated between mono- and dichloramine, Kelly and Sanderson noted that viral inactivation by chloramines proceeds more rapidly at pH 6-7 than at pH 8-10. This tendency indicates that dichloramine may be more virucidal than monochloramine since its proportion increases with increased hydrogen ion concentration.

In 1971, Chang proposed that 5.0 mg/liter dichloramine or 20 mg/liter monochloramine would be needed to inactivate enteroviruses by 99.99% in 10 min at 25 °C. Subsequently, Siders et al. presented evidence of the first comparative studies on viral inactivation due solely to monochloramine. At pH 9 at 15 °C, poliovirus 1 (Mahoney) was 10 times more resistant to monochloramine than was E. coli. Similarly, coxsackievirus A9 was approximately 4 times more resistant than E. coli. Siders› data can be compared to Chang›s theory on the disinfectant capacity of monochloramine. Assume that poliovirus inactivation has a temperature coefficient for a 10 °C rise (Q_{10}) of 3. Based on Chang's calculations that 20 mg/liter monochloramine at 25 °C would be needed to achieve 99.999% reduction of enterovirus in 10 min, one would extrapolate that 3 × 20 or 60 mg/liter monochloramine would be needed at 15 °C to achieve an equivalent reduction of poliovirus in 10 min. However, examination of the data of Siders et al. reveals that approximately 18 min were needed to achieve only 99% inactivation using 60 ppm monochloramine. Therefore, it appears that Chang slightly overestimated the virucidal capacity of monochloramine.

Efficacy against Parasites

Some studies, mostly in the field of wastewater treatment, have shown that ova and larvae of the helminth parasites that affect humans and that could occur in U.S. water supplies are resistant to current chlorination procedures. They can survive concentrations and exposure periods considerably in excess of those used in the treatment of municipal water supplies. In studies of various free-living nematodes, Chang *et al.* observed that 2.5 to 3.0 mg/liter of free chlorine for a 120-min contact period and 15 to 45 mg/liter of free chlorine for 1 min were not lethal. Free chlorine residuals as high as 95 to 100 mg/liter for 5 min killed only 40%-50% of the nematodes. Thus, it may be speculated that all the helminths, including their larvae, may approach the degree of resistance to chlorine that had been demonstrated by the free-living nematodes.

There have been a number of studies on the effectiveness of chlorine in destroying or inactivating cysts of the protozoan parasite, Entamoeba histolytica, in water, especially during the early 1940's. Varied results reflect primarily the different experimental conditions and techniques that were used. The presence of organic matter, pH, and temperature, as well as the concentration and form of chlorine and exposure period, have been shown to exert an influence on disinfection. However, the consensus is that, compared with bacteria, these cysts are rather resistant to current chlorination procedures, but are much less resistant than helminths.

Brady et al. conducted field-simulated studies with cysts in raw water that had been treated with calcium hypochlorite (CaOCl), resulting in pH levels ranging from approximately 7.5 to 8.0. They found that at temperatures of 23 °C-26 °C, exposures of 20 min and longer to residuals of 3 to 4 mg/liter were required to produce an estimated 99% cyst destruction as judged via a culture technique. Chang, also using a culture technique, studied the cysticidal effectiveness of calcium hypochlorite solution, chloramines, and gaseous chlorine in tap water as well as the effects of pH and organic matter on the biocidal activity. At contact periods of up to 30 min, gaseous chlorine was the most powerful, hypochlorite solution slightly less so, and chloramines the least. Increase in pH and organic matter reduced cysticidal efficacy. For comparison with Brady et al. , the ‹›lethal" residual concentration in tap water at 18 °C and pH 6.8-7.2 ranged from 2.8 to 3.2 mg/liter at 15 min, and from 1.8 to 2.2 mg/liter at 30 min.

Recently, Stringer et al. reported on comparative studies of the cysticidal efficacy of chlorine, bromine, and iodine as disinfectants. Using chlorine gas bubbled into buffered distilled water as stock, they obtained 99.9% cyst inactivation (as measured by excystment capability) after 15 min exposure to 2 mg/liter free chlorine in "clean water" at pH 6. However at pH 8 a contact time exceeding 60 min was required to achieve 99% mortality. In "secondary treated sewage effluent," Stringer et al. considered 13.7 mg/liter chlorine at pH 8 to be ineffectual as a cysticide.

In keeping with these findings, it is unlikely that the chlorine residuals generally maintained in distribution systems provide much protection against E. histolytica cysts in the event of contamination because of cross-connections, seepage, etc.

During the past 10 yr, a number of outbreaks of waterborne infections from Giardia lamblia (another intestinal protozoan) have been reported. Most incidents in the United States that were traced to municipal water supplies involved surface water sources where disinfection appeared to be the only treatment. The cysts of this parasite are thought to be as resistant to chlorine as those of E. hystolytica. However, there seem to be no studies of the resistance of this parasite to chlorine or other disinfectants.

Mechanism of Action

One of the earliest references to the mechanism of inactivation of microorganisms by chlorine resulted from the work of Chang. While studying the inactivation of E. histolytica cysts by chlorine, he observed greater uptake of chlorine and less survival at low pH than at high pH. This observation was associated with the increased inactivation efficiency of the undissociated hypochlorous acid. Supportive evidence for the hypothesis that permeability of the uncharged chlorine species is important in determining sensitivity to chlorine has been provided by Skvortsova and Lebedeva, Kaminski et al., and Dennis. Chang also noted that the inactivation of amoebic cysts was accompanied by microscopic damage to the cell nucleus, which was dependent on chlorine penetration.

The importance of penetration and/or damage to the permeability barrier of the cell membrane as a result of exposure to chlorine has been observed by several investigators. In 1945, Rahn suggested that the inactivation of bacteria by chlorine was due to multiple injuries to the cell surface. From their work with bacterial spores, Kulikovsky et al. implicated permeability damage as a mechanism of chlorine inactivation. Studies with Escherichia coli have shown that chlorine causes leakage of cytoplasmic material, first protein, then RNA and DNA, into the suspending menstruum. It also inhibits the biochemical activities that are associated with the bacterial cell membrane. Friberg observed that E. coli also loses nondialyzable phosphorus following exposure to chlorine. In a recent study, Haas demonstrated that chlorine caused certain bacteria and yeast to release organic matter or UV-absorbing material, presumably protein or nucleic acid or their precursors. This investigator also noted that chlorine affected the uptake and retention of potassium by these same microorganisms.

Green and Stumpf and Knox et al. indicated that destruction of bacteria by chlorine was caused by an inhibition of the mechanism of glucose oxidation. Specifically, they suggested that chlorine affected the aldolase enzyme of E. coli by oxidizing the sulfhydryl group that is associated with the enzyme. Venkobachar et al. recently reported that chlorine significantly inhibits both oxygen uptake and oxidative phosphorylation. The latter effect was attributed to inhibition of the respiratory enzyme rather than to a deficiency in phosphate uptake. However, it is unclear whether free or combined chlorine was used in these studies. Haas also observed chlorine to affect the respiration of bacteria as well as the rate of synthesis of protein and DNA. Others have also noted that chlorine affects the nucleic acids or physically damages DNA.

It appears that chlorine, having penetrated the cell wall, encounters the cell membrane and alters its permeability. Simultaneously or subsequently, the chlorine molecules may enter the cytoplasm and interfere with various enzymatic reactions. It should be noted that permeases and respiratory enzymes are associated with the cytoplasmic membrane of bacteria.

Chang supported the hypothesis that the rapid destruction of vegetative bacteria by chlorine was due to the extensive destruction of metabolic enzyme systems. He also addressed the subject of virus inactivation, commenting that viruses are generally more resistant to chlorine than bacteria. He associated this observation with the fact that viruses completely lack a metabolic enzyme system. He speculated that inactivation of viruses by chlorine probably result from the denaturation of the capsid protein. Furthermore, since protein denaturation is more difficult to achieve than destruction of enzymatic R—S—H bonds by oxidizing agents, it is understandable why greater levels of chlorine are required to inactivate viruses than bacteria. However, from their experimental work with the bacterial virus f2, Olivieri et al. concluded that chlorine caused initial lethal damage to the viral genome and that the capsid protein was affected after the virus was inactivated. Dennis reported that the incorporation of chlorine into the f2 bacterial virus is dependent on pH and that the higher rates of incorporation occur at lower pH values.

There is limited information in the literature on the mechanism of inactivation of microorganisms by chloramines. Nusbaum proposed that since low levels of inorganic chloramines were effective in inactivating bacteria, the mechanism of action must be essentially the same as that of hypochlorous acid on enzymes. Ingols et al. showed that monochloramine was not able to immediately and irreversibly oxidize sulfhydryl groups. Such oxidation would have resulted in the rapid inactivation of the bacteria. They hypothesized that since monochloramine required higher concentrations and longer contact times to destroy bacteria completely and could not readily and irreversibly oxidize the sulfhydryl groups of the glucose oxidation enzymes, its ability to inactivate microorganisms should be attributed to changes in enzymes that may not be involved in the inactivation of the organism by hypochlorous acid. Thus, while the sulfhydryl group may be the most vulnerable to a strong oxidant like hypochlorous acid, changes in other groups produced by the weaker oxidant, monochloramine, may lead also to microbial inactivation. More recent information indicates that the destructive effects of chloramine might be associated with the effects of chloramine on nucleic acids or DNA of cells.

Nusbaum suggested that the disinfective activity of dichloramine occurs by a mechanism similar to monochloramine, but there do not appear to be any data to support this contention.

Considering the mechanism of destruction or inactivation of microorganisms by chlorine and associated compounds, it is interesting to note that Fair et al. speculated that there might be three or four "targets" or points of attack and that perhaps all must be affected before there is death. This "multiple hit" concept supported the observation that monochloramine must alter groups other than the sulfhydryl group to be effective in the destruction of microorganisms.

Thus, the action of chlorine on microbes such as bacteria and amebic cysts may involve some or all of the steps in the following sequence: penetration of the disinfectant through the cell wall followed by attack on the cell membrane (the site of cellular respiration in bacteria) and disruption of permeability of the cell membrane, which leads to a loss of cell constituents, thereby disrupting metabolic functions within the cell including those involving nucleic acids. Changes in viability may result from this process. Experimental studies on virus demonstrated that chlorine caused initial damage on the viral nucleic acid while leaving the capsid protein unaffected until after the virus was inactivated.

Chlorine is the most widely used water supply disinfectant in the United States. Depending upon the predominant species of chlorine, hypochlorous acid, and/or hypochlorite ion, disinfection with chlorine can achieve greater than 99.9% destruction of bacteria. For example, a chlorine residual of 0.2 to 1.0 mg/liter and a contact time of 15 to 30 min will inactivate 99.9% of E. coli. According to Walton, a properly designed, constructed, and operated water treatment plant, consisting of chemical coagulation, sedimentation, filtration, and disinfection, can remove or destroy more than 99.999% of the coliform bacteria that are present. Although most investigations on the removal or destruction of bacteria have used E. coli, there is evidence that the bacterial pathogens, e.g., Salmonella typhi, respond somewhat similarly to E. coli.

Laboratory studies have demonstrated that that there is limited virus inactivation after the added chlorine has reacted with any ammonia that is in the water. Most inactivation probably occurs in the first few seconds before the chlorine has completed its reaction with ammonia.

Recent reports of enhanced chlorine resistance of certain viral and bacterial strains should be investigated and the mechanism of increased resistance elucidated, if the reports are corroborated.

Other recommendations, applicable to other agents as well as to chlorine, are included after the evaluations of the other methods of disinfection.

Ozone

Chemistry of Ozone in Water

Ozone has the molecular formula O_3, a molecular weight of 48 g/mol, and a density, as a gas, of 2.154 g/liter at 0 °C and 1 atm. It is approximately 13 times more soluble in water than is oxygen. At saturation in water at 20 °C, a 2% weight mixture of ozone and oxygen contains about 11 mg of ozone and 40 mg of oxygen per liter.

Ozone has a half-life in pure distilled water of approximately 40 min at pH 7.6, but this decreases to 10 min at pH 8.5 at 14.6 °C. Rising temperatures increase the rate of decomposition. Because its half-life is so short, ozone must be generated on the site where it is to be used.

Ozone is a powerful oxidant that reacts rapidly with most organic and many inorganic compounds. It does not convert chloride to chlorine under test conditions (U.S. Environmental Protection Agency, 1976), nor does it react extensively with ammonia (NH_3). However, bromide and iodide are oxidized to bromine and iodine. Singer and Zilli reported that oxidation of ammonia was pH-dependent. At pH 7.0, 9% of a 29 mg/liter ammonia nitrogen $(NH_3 - N)$ solution was oxidized in 30 min, but at pH 9.0 70% of a 24.4 mg/liter solution was oxidized in the same time. During disinfection, only minor amounts of ammonia are oxidized when ozone is used. Ozone's limited reaction with ammonia is desirable, but its fast reaction rate with most organic and many inorganic compounds further shortens its persistence in water.

Production and Application of Ozone

Ozone is produced on site from a stream of clean dry air or oxygen by passing an electrical discharge between electrodes that are separated by a dielectric. Approximately twice the percent of ozone by weight is obtained if oxygen, rather than air, is used as the feed stream. Power requirements are about 13 to 22 kWh/kg of ozone that is generated from air and approximately half that when oxygen is used. Compressors and dryers may increase these requirements by 20% to 50%. Other factors affecting efficiency are the rate of gas flow, applied voltage, and the temperature of the gas. The heat that is produced during the process must be removed by cooling with either air or water.

The ozone gas stream must be fed into the water to effect the transfer of ozone. The usual methods are to inject the ozone gas stream through an orifice at the bottom of a co- or countercurrent contact chamber or to aspirate the gas into a contact chamber where it is mixed with the water mechanically. Successful design and operation of the contactor system is necessary to minimize costs of the operation.

Commercial equipment is available in a wide range of capacities—from a few grams of ozone per day to more than 40 kg/day. Larger capacities are obtained by adding additional units. Successful delivery of ozone to the water to be treated requires a dependable power supply and reasonably maintenance-free ozonization equipment.

Ozone has been used in a great number of water treatment plants throughout the world. However, in small institutions and private residences, its use appears limited, because it requires dependable power supplies and, usually, a second disinfectant to furnish a disinfecting residual in the system. The maintenance and repairs that are required for the specialized ozone generation equipment provide further barriers against the use of ozone by small institutions.

Analytical Methods

The disinfection process is usually controlled in one of two ways: by the dosage of a specified amount of ozone or by the maintenance of a specified minimum residual for a given time. Residual measurements in both the gas stream and water are sometimes required. Standard Methods contains descriptions of the measurement of ozone in water by the iodometric, orthotolidine-manganese sulfate, and orthotolidine-arsenite methods. Of these methods the iodometric method, which is subject to the fewest interferences, is the method of choice. Determinations must be made immediately since ozone decomposes rapidly.

In all three methods, the oxidant compounds that result from the reaction of ozone with contaminants in water may react with the test reagents, thereby indicating a higher concentration of ozone than is actually present. This is particularly true in the presence of organic matter, which results in the formation of organic peroxides. In the iodometric method this interference and others are minimized by stripping the ozone from the sample with nitrogen or air and absorbing it from this gas stream in an iodide solution.

Schecter developed a UV spectrophotometric method to measure the triiodide that is formed by the oxidation of iodide by ozone. She reported a better sensitivity at low ozone concentrations (0.01 to 0.3 mg/liter) than achieved with the normal titration method. The effects of interferences on the direct measurement of ozone without sparging the ozone to a separate iodide solution were not indicated.

Analytical determination of ozone in water in the presence of other oxidants is poor. Considerable work in this area is needed.

Residual

As discussed above, ozone is unstable in water with a half-life of approximately 40 min at pH 7.6 and 14.6 °C. Many regard the half-life in water supplies at higher ambient temperatures to be 10 to 20 min.

Peleg concluded that the possible species to be found in an ozone solution were ozone, hydroxyl radicals $(\cdot OH)$, hydroperoxyl radicals $(HO_2 \cdot)$, oxide radicals $(O \cdot)$, ozonide radicals $(O_3 \cdot)$, and, possibly, free oxygen atoms $(\cdot O \cdot)$. Hydrogen peroxide (H_2O_2) may also be present by dimerization of the hydroxyl radicals. There have been no studies on the disinfecting activities of these individual species except for those on hydrogen peroxide, which is a poor biocide (when compared to chlorine). Peleg concluded that evidence indicates that the dissociation species are better disinfectants than ozone.

Biocidal Activity

Efficacy against Bacteria

Wuhrmann and Meyrath found that 99% of Escherchia coli were inactivated in 21 s with 0.0125 mg/liter residual ozone, in 62 s with 0.0023 mg/liter residual ozone, and in 100 s with 0.0006 mg/liter residual ozone at pH 7.0 and 12 °C. Spores of a Bacillus species were much more resistant, requiring 2 min with 0.191 mg/liter residual ozone and 5 min with 0.049 mg/liter residual ozone for 99% inactivation. These data were obtained at pH 7.2 and 22 °C. The observed ozone residuals were reported as being constant throughout the test periods.

Katzenelson et al. found that an initial residual ozone concentration of 0.04 mg/liter in demand-free water inactivated approximately 3 logs of E. coli in 50 s, while a concentration of 1.3 mg/liter achieved the same degree of inactivation in 5 s. They obtained their inactivation data from experiments in which the loss of initial ozone residuals did not exceed 20%. Their work was conducted at 1 °C and pH 7.2. The ozone was determined by the method of Schecter.

Using washed cells of E. coli, Bacillus megaterium, and Bacillus cereus, which were suspended in deionized water with 106 cells/ml, and different concentrations of ozone up to 0.71 mg/liter, Broadwater et al. demonstrated that the inactivation of all three test organisms gave an "all-or-none" response. They measured the ozone by stripping it into iodide at the end of the contact period. Consequently, any demand should have used up whatever ozone was needed. They did measure what they presumed to be ozone and not some breakdown product. Thus, they appear to have eliminated the ozone demand problem as much as is possible with present techniques. However, it is possible that their results reflect some effect or artifact not yet understood. With the constant contact time of 5 min, no inactivation of vegetative cells of E. coli or B. megaterium occurred until the initial residual concentration of ozone was 0.19 mg/liter, when the density of viable cells of both species decreased to "near zero." With the vegetative cells of B. cereus, an initial ozone residual of 0.12 mg/liter was required before any inactivation occurred, and then it was nearly complete. Spores of B. megaterium and B. cereus were not inactivated until the ozone residual reached approximately 2.29 mg/liter, when the spores of both organisms showed an "all-or-none" response. All of the inactivation experiments were performed at 28 °C, but the pH was not reported.

Burleson et al. determined that E. coli, Staphylococcus aureus, Salmonella typhimurium, Shigella flexneri, Psuedomonas fluorescens, and Vibrio cholerae were all reduced by 7.5 logs in 15 s after exposure to approximately 0.5 mg/liter of ozone at 25 °C in phosphate-buffered saline. The pH was not reported. In this study, the bacteria were placed in unozonized water (no initial residual), and the ozone was then sparged into the water. After 15 s the ozone concentration in solution reached approximately 0.5 mg/liter. This technique does not render quantitative data. Ozone was determined by spectrophotometric measurement of iodine that was released from iodide without stripping of the ozone.

The inactivation of E. coli with initial counts of about $5' \ 10^5$/ml in buffered demand-free water at 11 °C was measured by Ross et al. For 99% inactivation with 0.1 mg/liter of initial ozone residuals, contact times of 16.5 and 21 s were required at pH 6 and 10, respectively. Ozone was measured by the diethyl-p-phenylenediamine (DPD) method.

Farooq and Farooq et al. studied the effects of pH and temperature in ozone inactivation studies with Mycobacterium fortuitum and Candida parapsilosis (a yeast). They concluded that if the ozone residual remains constant, the disinfection capability will not be affected by a change in pH. They also demonstrated that for a given dosage a rise in temperature increases the rate of inactivation, even though the ozone residual was decreased. (Ozone was less soluble at higher temperature). Vorchinskii concluded that whereas the bactericidal dose of ozone at 4 °C to 6°C was unity, the corresponding dose at 18 °C to 21 °C was 1.6 and at 36 °C to 38 °C was 3.6.

The work of Farooq and his colleagues is in agreement with that of Morris, who observed that the disinfection capability of ozone does not change significantly with pH, at least over the normal pH range (6 to 8.5) of water supplies.

Efficacy against Viruses

Katzenelson et al. investigated the inactivation of poliovirus 1 at 5 °C and pH 7.2. A 99% inactivation of the virus occurred in less than 8 s with 0.3 mg/liter of initial residual ozone. Their experimental methods were the same as those described above for bacteria. They also observed that inactivation resulted from two distinct stages (rates) of action. The first stage of inactivation, less than 8 s long, produced a virus inactivation of from 99% to 99.5%, depending upon the ozone residual. The second stage, which lasted from 1 to 5 min, still left some viruses infective. Additional work showed that the slower second stage inactivation apparently involved the inactivation of viruses that were clumped together. The single virus particles were inactivated during the first stage. After ultrasonic treatment, 99.5% of the virus was inactivated within 8 s and more than 99.99% within 3 min at an initial ozone residual of 0.1 mg/liter.

Coin et al. investigated the inactivation of poliovirus 1 (Mahoney) in a batch system in which a measured amount of ozone was introduced into a virus suspension in distilled and filtered river water. In distilled water, nearly 1 log of the virus was inactivated in 4 min at a 4-min ozone residual of approximately 0.23 mg/liter. In river water, the inactivation was in excess of 99.99% in 4 min when the ozone concentration, also measured at 4 min, exceeded 0.3 mg/liter. The ozone was measured by the iodide titration without stripping of the ozone. Rather large ozone demands existed in this system. An initial 5 mg/liter residual ozone concentration in the distilled water decreased to about 0.6 to 0.8 mg/liter in 4 min. The same initial residual in river water decreased to between 0.2 to 0.6 mg/liter. Temperature and pH were not reported.

Keller et al. studied ozone inactivation of viruses in a water supply by using both batch tests and a pilot plant with a 38 liter/min (10 gal/min) flow rate. Inactivation of poliovirus 2 and coxsackievirus B3 in 5 min was greater than 99.9% in the batch tests when the ozone residual was between 0.8 and 1.7 mg/liter at the end of the 5-min contact period. Initial ozone residuals varied from 1.6 to 2.8 mg/liter. At the pilot plant, greater than 99.999% inactivation of coxsackievirus B3 could be achieved with an ozone dosage of 1.45 mg/liter, which provided an ozone residual of 0.28 mg/liter after 1 min in lake water. Temperature and pH conditions were not reported. Ozone was measured by the iodide method without prior stripping of the ozone.

Burleson et al. obtained inactivations of > 99.9% for vesicular stomatitis virus, > 99.99% for encephalomyocarditis virus, and 99.99% for GD VII virus in 15 s with an ozone residual of approximately 0.5 mg/liter of ozone. The experiment was conducted in the same manner as that described above for their studies with bacteria. Evison reported data for the inactivation of a number of viruses in buffered water at pH 7.0 and 25 °C.

Table: Ozone Required for Inactivation of Viruses in 10 Minutes at pH 7.0 and 25 °C.

Virus	Ozone Required, mg/liter	
	99.9% Inactivation	99% Inactivation
Coxsackie B3	0.6	0.095
Polio 3	0.22	0.082
Polio 1	0.095	0.042
Echo 1	0.086	0.044
Coxsackie B5	0.076	0.053
Polio 2	0.052	0.039

The reported ozone concentrations were evidently those measured initially and maintained by the addition of ozone during the experiment. Ozone was measured by a colorimetric version of Palin's DPD technique. The Evison data show that more ozone or a longer time are required for inactivation than do the data of other workers. This may have resulted from the virus purification used. Her viruses were purified by low-speed centrifugation and filtration through an 0.2-μm membrane filter. These cleanup procedures are neither as complete nor as thorough as those used by other investigators. The unremoved cell debris and organic matter offer protection to the virus. Either higher ozone residuals or longer contact times would be required to inactivate such preparations to the same extent as clean virus. Other data showed that the inactivation of coliphage 185 was relatively unaffected by pH's ranging from 6.0 to 8.0 at ozone concentrations exceeding 0.05 to 0.10 mg/liter. Evison also concluded that the rate of inactivation of the coliphage 185 by ozone was much less affected by temperature than was the inactivation by chlorine.

Farooq observed a 99% inactivation of 10^5/ml poliovirus 1 in 30 s in distilled water at an initial ozone residual concentration between 0.23 and 0.26 mg/liter at 24 °C and pH 7.0. In this study, ozone was measured by the Schecter method.

Sproul et al. reported total inactivation in 10 s of poliovirus 1 (Sabin), which had initial concentrations of 1.3×10^2 to 2.4×10^3 plaque-forming units (PFU)/ml. The initial ozone residual of 0.012 to 0.085 mg/liter decreased by approximately one-third in 40s. The experiments were conducted with the Sharpe dynamic reactor. Ozone was measured by the Schecter method.

Efficacy against Parasites

Ozone may have application as an antiparasitic agent in the treatment of water supplies but only limited information is available. Newton and Jones reported that ozone, with 5-min residuals as low as 0.3

mg/liter, inactivated from 98% to >99% of Entamoeba histolytica cysts that were suspended in water. Initial ozone residuals that were required to obtain 5-min residuals of 0.3 mg/liter varied from 0.7 mg/liter to 0.9 mg/liter. With the ozone concentrations used, the cysticidal action was not affected by temperatures from 10 °C to 27 °C nor by pH's between 6.5 to 8.0. Ozone was measured by titration of iodine, which was released from iodide directly in the reactors without removal of ozone by sparging.

Mechanism of Action

Investigations of the inactivation of bacteria by ozone have centered on the action of ozone on the cell membrane. Scott and Lesher concluded from their work with E. coli that the primary attack of ozone occurred on the double bonds of the fatty acids in the cell wall and membrane and that there was a consequent change in the cell permeability. Cell contents then leaked into the water. This was confirmed by Smith. Prat et al., examining extracts from ozonized E. coli, reported that thymine was more sensitive to ozone attack than were cytosine and uracil.

Riesser et al. showed that ozone attacked the protein capsid of poliovirus 2 in such a manner that the virus was not taken up into susceptible cells. An electrophoretic study showed complete loss of viral proteins in a poliovirus 2 sample that had showed an inactivation of 7 logs in 20 min.

Inactivation with ozone at specified ozone residuals is relatively insensitive to pH's between 6.0 and 8.5. Moreover, ozone does not react with ammonia over this same range when short detention times are used. The data on temperature are not sufficiently firm to permit conclusions concerning its effect on disinfection. Ozone must be generated on site, and the process is relatively energy intensive. To make economic comparisons of ozone with other disinfectants, the cost of local power must be ascertained.

Available kinetic data on ozone inactivation are presented in Table. The c · t product for 99% inactivation of E. coli appears to be approximately 0.006 at near-neutral pH values and at temperatures of ≤10 °C. There was less consistency for the c · t products of poliovirus 1. These values varied from < 0.005 to 0.42 at pH 7 and at temperatures from 5 °C to 25 °C.

Table: Concentration of Ozone and Contact Time Necessary for 99% Inactivation of Escherichia coli and Polio 1 Virus.

Test Microorganism	Ozone, mg/liter	Contact Time, min	c·t	pH	Temperature, °C
E. coli	0.07	0.083	0.006	7.2	1
	0.065	0.33	0.022	7.2	1
	0.04	0.50	0.02	7.2	1
	0.01	0.275	0.027	6.0	11
	0.01	0.35	0.035	6.0	11
	0.0006	1.7	0.001	7.0	12
	0.0023	1.03	0.002	7.0	12
	0.0125	0.33	0.004	7.0	12
Polio 1	< 0.3	0.13	< 0.04	7.2	5
	0.245	0.50	0.12	7.0	24
	0.042	10	0.42	7.0	25
	< 0.03	0.16	< 0.005	7.0	20

The $c \cdot t$ products vary over a broad range. These variations illustrate the difficulty of doing quantitative experimentation with ozone and microorganisms in water. Among other reasons, these difficulties are caused by undetected ozone demand in the water, poor analytical techniques for residual ozone, and nonuniformity of microorganisms from one laboratory to another. As an example of the latter, different strains of poliovirus with different inactivation rates are used, but the inactivation data are frequently not reported as strain-specific. Furthermore, viruses frequently exist in an undetected clumped state rather than in the presumed single discrete particle state. Higher $c \cdot t$ products are required for the clumped viruses than for the unclumped ones.

Because of ozone's relatively short half-life in water, another disinfectant must be added to maintain a disinfection capability in the distribution system. The most effective disinfectant, its optimum concentration, and method of addition must be determined. The disinfection process with ozone will probably be controlled by specifying the ozone residual at the beginning and end of a given contact time.

Future research studies with ozone should be conducted to:

- Provide data on the inactivation of enteric pathogens;

- Provide more data on the inactivation of protozoan cysts, especially those of giardia lamblia;

- Provide analytical methods that are specific for ozone;

- Provide additional definitive data on bacteria and viruses, and eliminate discrepancies;

- Provide operating data from full-scale drinking water plants to demonstrate reliability of operation, operating costs, ozone dosages, residuals, contact times, and disinfection results.

Chlorine Dioxide

Chlorine dioxide (ClO_2) was first prepared in the early nineteenth century by Sir Humphrey Davey. By combining potassium chlorate $(KClO_3)$ and hydrochloric acid (HCl), he produced a greenish-yellow gas, which he named "euchlorine." Later, this gas was found to be a mixture of chlorine dioxide and chlorine. The bleaching action of chlorine dioxide on wood pulp was recognized by Watt and Burgess.

Large quantities of chlorine dioxide are produced each day in the United States. Although its primary application has been the bleaching of wood pulp, it is also used extensively for bleaching and dye stripping in the textile industry and for bleaching flour, fats, oils, and waxes.

In the United States, chlorine dioxide was first used in 1944 at the water treatment plant in Niagara Falls, New York, to control phenolic tastes and odors arising from the presence of industrial wastes, algae, and decaying vegetation. Granstrom and Lee surveyed water treatment plants believed to be using chlorine dioxide.

The majority of respondents (956 plants) were using it for taste and odor control. Other uses reported were algal control (7 plants), iron and manganese removal (3 plants), and disinfection (15 plants).

Sussman compiled a partial listing of plants using chlorine dioxide. He reported that the compound is used primarily to control taste and odor in the United States. In England, Italy, and Switzerland, it is used for disinfection of water supplies.

The Chemistry of Chlorine Dioxide in Water

Chlorine dioxide reacts with a wide variety of organic and inorganic chemicals under conditions that are usually found in water treatment systems. However, two important reactions do not occur. Chlorine dioxide per se does not react to cause the formation of trihalomethanes (THM's). However, THM's will be formed if the chlorine dioxide is contaminated with chlorine. Such a situation may occur when chlorine is used in the preparation of chlorine dioxide.

Chlorine dioxide does not react with ammonia, but will react with other amines. The amine structure determines reactivity. Tertiary amines are more reactive with chlorine dioxide than secondary amines, which, in turn, are more reactive than primary amines.

Production and Application

Chlorine dioxide condenses to form an unstable liquid. Both the gas and liquid are sensitive to temperature, pressure, and light. At concentrations above 10% in air, chlorine dioxide may be explosive, and at 4% in air, it can be detonated by sparks. As a result, the preparation and distribution of chlorine dioxide in bulk have not been deemed practical. It has been generated and used on site.

Sodium chlorate $(NaClO_3)$ or sodium chlorite $(NaClO_2)$ may be used to generate chlorine dioxide. The method of production will depend upon the amount of chlorine dioxide that is required. The reduction of sodium chlorate is the more efficient process and is generally used when large volumes and high concentrations of chlorine dioxide are needed. Commercial processes that are used in North America for large-scale production of chlorine dioxide are based on the three reactions listed below. To reduce the sodium chlorate, each process uses a different agent: sulfur dioxide (SO_2), methanol (CH_3OH), and the chloride ion (Cl^-).

All of these processes are used in the pulp and paper industry. They can also be used to prepare chlorine dioxide for the large waterworks that might require several metric tons per day. Small quantities of chlorine are formed during the side reactions and intermediate reactions in these processes. A more detailed review of the chemistry that is involved in the production of chlorine dioxide from chlorate is given by Gall and Gordon et al.

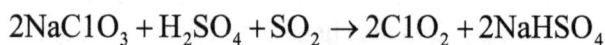

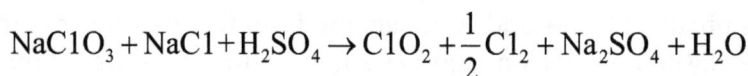

$$2NaClO_3 + H_2SO_4 + SO_2 \rightarrow 2ClO_2 + 2NaHSO_4$$

$$2NaClO_3 + CH_3OH + H_2SO_4 \rightarrow 2ClO_2 + HCNO + Na_2SO_4 + 2H_2O$$

$$NaClO_3 + NaCl + H_2SO_4 \rightarrow ClO_2 + \frac{1}{2}Cl_2 + Na_2SO_4 + H_2O$$

Chlorine dioxide can be prepared from chlorine and sodium chlorite through the following reactions:

$$Cl_2 + H_2O \rightarrow HOCl + HCl$$

$$HOCl + HCl + 2NaClO_2 \rightarrow 2ClO_2 + 2NaCl + H_2O$$

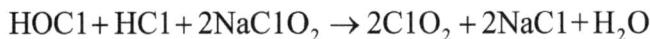

The theoretical weight ratio of sodium chlorite to chlorine is 1.00:0.39. With available sodium chlorite (80%), the weight ratio is 1:0.30. In practice, Gall recommended a chlorite to chlorine ratio of 1:1. The excess chlorine lowers the pH, thereby increasing the reaction rate and optimizing the yield of chlorine dioxide. Dowling reported that the maximum theoretical yield of chlorine dioxide was produced when the ratio was normally maintained at a minimum of 1.0 : 0.5.

Alternatively, chlorine dioxide may be prepared from sodium hypochlorite (NaOCl) and sodium chlorite. The sodium hypochlorite is acidified to yield hypochlorous acid (HOCl), and the chlorine dioxide is generated according to reaction above. Each of the methods produces a solution containing both chlorine and chlorine dioxide.

Chlorine dioxide may also be prepared by the addition of a strong acid, such as sulfuric acid (H_2SO_4) or hydrochloric acid, to sodium chlorite as shown in the following reactions:

$$10NaClO_2 + 5H_2SO_4 \rightarrow 8ClO_2 + 2Na_2SO_4 + 2HCl + 4H_2O$$

$$5NaClO_2 + 4HCl \rightarrow 4ClO_2 + 5NaCl + 2H_2O$$

Although some investigators have claimed that this method produces chlorine-free chlorine dioxide, Feuss and Schilling reported that chlorine is also formed. Dowling indicated that chlorine was formed even when sulfuric acid was used.

Analytical Methods

Chlorine dioxide is one of the few stable nonmetallic inorganic free radicals. It does not contain available chlorine in the form of hypochlorous acid or hypochlorite ion (OCl^-). However, concentrations of chlorine dioxide are often reported in terms of available chlorine. The chlorine atom in chlorine dioxide has a valance of + 4. A reduction to chloride results in a gain of five electrons. In terms of available chlorine, chlorine dioxide has 263% or more than 2.5 times the oxidizing capacity of chlorine.

$$\frac{5e^- \times 35.45}{67.45} \times 100 = 263\%$$

The weight ratio of chlorine dioxide to available chlorine is 67.45 to 35.45 or 1.9. However, in water treatment practices this increased oxidizing capacity is rarely realized. The reduction of chlorine dioxide depends heavily on pH and the nature of the reducing agent. At neutral or alkaline pH, chlorine dioxide is reduced to chlorite, a net gain of one electron. Thus, only one-fifth or 20% of its oxidizing capacity is utilized. At low pH, the chlorite (ClO_2^-) is reduced to chloride (Cl^-) releasing the remaining four available electrons.

Chlorine dioxide may be determined iodometrically, amperometrically, spectrophotometrically, and colorimetrically.

Several studies contain comparisons of various analytical methods and procedures for the measurement of chlorine dioxide. Adams *et al.* compared the H-acid, tyrosine, amperometric, and diethyl-*p*-phenylenediamine (DPD) procedures for free chlorine, chlorine dioxide, and chlorite. They reported DPD to be the most reliable, as did Dowling, Miltner, and the U.S. Public Health Service. Myhrstad and Samdal noted that the DPD method yielded consistently higher residual measurements for chlorine dioxide than those that are produced with other analytical methods. After analysis with acid chrome violet K, chlorine dioxide was not observed in the water of the distribution system; however, chlorite was found. The residuals that were previously interpreted as chlorine dioxide were apparently due to chlorite.

Recently, more sophisticated procedures were suggested for the determination of chlorine dioxide. Moffa et al. reported the use of electron spin resonance, and Issacsson and Wettermark described a chemiluminescent method for active chlorine compounds. Stevenson et al. reviewed electrochemical methods and presented preliminary results for a membrane amperometric probe. Under development is a sensor that shows a linear response region from about 0.5 to 10 mg/liter. The response to hypochlorous acid and chloramines was low, and the sensor does not measure chlorite or other ionized species.

No one procedure appears to possess the necessary sensitivity, selectivity, and simplicity to permit reliable determinations in the treatment of water. Each of the titration methods are prone to error because of volatilization. They are time-consuming and particularly complex when differentiation of chlorine and oxychloro species are necessary. The colorimetric procedures require strict control of pH, temperature, and reaction times and will be affected by turbidity. In addition, the selectivity of the indicators for chlorine dioxide is questionable. The direct spectrophotometric determination of chlorine dioxide at 360 nm is selective and rapid but is not sufficiently sensitive for use in water. Limited experience with the more recent procedures (chemiluminescence and membrane amperometric probe) does not permit an evaluation.

In practice, the principal distinction that must be made is that between the active biocidal species (hypochlorous acid, the hypochloride ion, and chlorine dioxide), the moderately biocidal species (monochloramine $[NH_2Cl]$, dichloramine $[NHCl_2]$, and nitrogen trichloride $[NCl_3]$), and the relatively nonbiocidal species (chlorite and chlorate $[ClO_3^-]$ ions). This is imperative when the primary purpose for the addition of chlorine dioxide is the inactivation of microorganisms. Furthermore, when the formation of THM's is to be considered, the distinction between free chlorine and chlorine dioxide becomes important.

Biocidal Activity

Efficacy against Bacteria

Experimental data on the efficacy of chlorine dioxide as a disinfectant became available in the early 1940's. McCarthy reported that chlorine dioxide was an effective bactericide in water with a low organic content. When the levels of organic material in the water were high, chlorine dioxide was less effective.

Ridenour and Ingols reported that chlorine dioxide was at least as effective as chlorine against Escherichia coli after 30 min at similar residual concentrations. Both chlorine and chlorine dioxide residues were determined by the orthotolidine-arsenite (OTA) method. The bactericidal activity of chlorine dioxide was not affected by pH values from 6.0 to 10.0. Ridenour and Armbruster extended their observation to other enteric bacteria. The common waterborne pathogens were similarly inactivated with chlorine dioxide. They also reported that the efficiency of chlorine dioxide decreased as the temperature decreased from 20 °C to 5 °C.

Ridenour et al. found chlorine dioxide to be more effective than chlorine (based on OTA residuals) against bacterial endospores. They indicated that less weight of chlorine dioxide than chlorine is required to inactivate the spores of Bacillus mesentericus in either demand-free water or in waters containing ammonia. In the waters containing ammonia, chlorine had to be applied beyond breakpoint before efficient sporicidal activity was observed.

The work of Ridenour and colleagues is not discussed in depth because the small amounts of free chlorine that are produced during the generation of chlorine dioxide are not distinguished from the chlorine dioxide by the OTA method that they used for both stock solutions and residual determination. In their 1947 paper, the survival measurements were ± and not quantitative. In their 1949 paper, the survival measurement was quantitative, but only one contact time was used, 5.0 min. Since chlorine dioxide is a rapidly acting disinfectant, $c \cdot t$ products may be misleading using this contact time.

Russian investigators found chlorine dioxide to be a more effective or at least as effective a bactericide as chlorine. Additional data were reported for Bacillus anthracis. They observed that the efficiency of chlorine dioxide decreased as the pH of the system containing the B. anthracis increased.

Early studies on disinfectant activity are difficult to interpret because the methods of preparing chlorine dioxide invariably included the addition or production of chlorine. Analytical procedures were not sufficiently advanced to differentiate between chlorine dioxide and other oxychloro species. Thus, the quantitative analyses of stock solutions and reports of dose and residual chlorine dioxide may be in error. This suggests that the initial and residual concentrations of chlorine dioxide were probably lower than reported values and that the comparative bactericidal efficiency would suffer accordingly. In addition, the older investigations did not take into account the volatility of chlorine dioxide. Depending upon concentration and length of exposure, losses of 7% to 30% can occur within an hour.

Many of the difficulties that were encountered during the early studies were overcome in a series of studies reported by Benarde and co-workers during the mid-1960's. Their work on disinfection was based heavily on the improved methods of preparing and analyzing chlorine dioxide, which were reported by Granstrom and Lee. They prepared chlorine dioxide by oxidizing sodium chlorite with persulfate $\left(S_2O_8^{2-}\right)$ under acid conditions. The resulting chlorine dioxide was swept to a collection vessel by high purity nitrogen gas. Chlorine dioxide was measured spectrophotometrically at 357 nm.

Bernarde et al. compared the bactericidal effectiveness of chlorine with that of chlorine dioxide at pH 6.5 and 8.5 in a demand-free buffered system. At pH 6.5, both chlorine and chlorine dioxide inactivated a freshly isolated strain of E. coli in less than 60 s. Chlorine was slightly more effective at the lower dosages at the lower pH. At pH 8.5, chlorine dioxide was dramatically more effective than chlorine. Greater than 99% inactivation of E. coli was observed in 15 s with 0.25 mg/liter chlorine dioxide, while, under the same conditions, chlorine required almost 5 min for similar inactivation. Chlorine dioxide was significantly more efficient than chlorine in the presence of high levels of organic and nitrogenous material. Bernarde et al. also reported that temperature affects the rate of inactivation of bacteria with chlorine dioxide. A decrease in disinfectant activity was observed as temperature decreased from 30 °C to 5 °C. For 99% inactivation of E. coli with 0.25 mg/liter of chlorine dioxide, 190, 74, 41, and 16 s were required at 5 °C, 10 °C, 20 °C, 30 °C, respectively. More recent work by Cronier in a clean system also demonstrated the excellent bactericidal activity of chlorine dioxide. Results of both studies are shown in table.

Table: Concentrations of Chlorine Dioxide and Contact Times Necessary for 99% Inactivation of Escherichia coli.

Test Microorganism	Chlorine Dioxide, mg/liter	Contact Time, min	c·t	pH	Temperature, °C
E. coli (freshly isolated from feces)	0.25	1.8	0.45	6.5	5
	0.50	0.83	0.41	6.5	5
	0.75	0.50	0.38	6.5	5
	0.25	1.2	0.30	6.5	10
	0.50	0.47	0.24	6.5	10
	0.75	0.3	0.23	6.5	10
	0.25	0.68	0.17	6.5	20
	0.50	0.35	0.18	6.5	20
	0.75	0.25	0.19	6.5	20
	0.25	0.27	0.07	6.5	32
	0.50	0.22	0.11	6.5	32
	0.75	0.15	0.11	6.5	32
E. coli (ATCC 11229)	0.30	1.8	0.54	7.0	5
	0.50	0.98	0.49	7.0	5
	0.80	0.58	0.41	7.0	5
	0.30	1.3	0.39	7.0	15
	0.50	0.75	0.38	7.0	15
	0.80	0.47	0.38	7.0	15
	0.30	0.98	0.29	7.0	25
	0.50	0.55	0.28	7.0	25
	0.80	0.35	0.28	7.0	25

Efficacy against Viruses

In the mid-1940's, there were also investigations on the virucidal activity of chlorine dioxide. Ridenour and Ingols reported that chlorine dioxide was as effective as chlorine against a mouse-adapted strain of poliovirus (Lansing). Again, their comparison was based upon OTA-determined residual levels of each disinfectant. Hettche and Ehlbeck found chlorine dioxide to be more effective against poliovirus than either chlorine or ozone. In addition to the difficulties that are associated with the preparation and measurement of chlorine dioxide and chlorine, the early virus studies were also saddled with difficult and time-consuming virus test systems.

The first definitive work on the virucidal activity of chlorine dioxide was done in Göteborg, Sweden. Warriner showed that the rate of inactivation of poliovirus 3 increased with increasing pH at pH values of 5.6 to 8.5. Similar to the action on bacteria, the viral inactivation occurred rapidly. When chlorine and chlorine dioxide were combined, the inactivation was synergistic. The inactivation with the two chlorine species together was more efficient at lower pH, but the presence of chlorine dioxide enhanced inactivation by chlorine at the higher pH.

Table: Concentrations of Chlorine Dioxide and Contact Times Necessary for 99% Inactivation of Viruses.

Test Microorganism	Chlorine Dioxide, mg/liter	Contact Time, min	c·t	pH	Temperature, °C
Poliovirus 3	0.5	5.0	2.5	5.6	20
	0.5	5.0	2.5	7.2	20
	0.5	0.25	0.125	8.5	20
	1.6	5.0	8.00	5.6	20
	1.6	1.0	1.60	7.1	20
	1.6	0.25	0.40	8.0	20
Poliovirus 1	0.3	16.6	5.0	7.0	5
	0.5	12.0	6.0	7.0	5
	0.8	6.8	5.4	7.0	5
	0.3	4.2	1.3	7.0	15
	0.5	2.5	1.25	7.0	15
	0.8	1.7	1.4	7.0	15
	0.3	3.6	1.08	7.0	25
	0.5	2.0	1.0	7.0	25
	0.8	1.5	1.2	7.0	25
Coxsackievirus A9	0.3	1.2	0.4	7.0	15
	0.5	0.05	0.34	7.0	15
	0.8	0.25	0.20	7.0	15

Cronier compared the inactivation of E. coli (ATCC 11229), poliovirus 1, and coxsackievirus A9. All of the microorganisms that were tested were sensitive to low concentrations (< 1 mg/liter) of chlorine dioxide. Poliovirus 1 and coxsackievirus A9 were more resistant than E. coli. Similar to its bactericidal activity, chlorine dioxide was more effective as a virucide at higher pH. Cronier also reported that on a weight basis, it was similar to hypochlorous acid and better than hypochlorite ion, monochloramine, and dichloramine.

The effect of bentonite $\left(AlO_3 \cdot 4SiO_2 \cdot H_2O \right)$ (added to the test as a model for turbidity) on disinfection with chlorine dioxide was studied by Scarpino et al. At turbidity levels below 2.29 nephelometric turbidity units (NTU) at 25 °C, no protection was afforded to poliovirus 1 by the bentonite. However, at 3.22 and 14.1 NTU›s, poliovirus inactivation was noticeably decreased. The bentonite at these levels appeared to offer protection to the virus.

Efficacy against Parasites

The data that are available on the efficacy of chlorine dioxide on helminths or protozoan cysts do not appear to be suitable for comparison with the action of other disinfectants.

Comparative Biocidal Activity

The data presented in previous two tables were collected for demand-free systems within the last 15 yr, when relatively reliable chemical methods and/or quantitative biocidal assay procedures were used. Even at 5°C and chlorine dioxide concentrations of 0.25 to 0.30 mg/liter, less than 2 min was required for 99% inactivation. Increases in chlorine dioxide concentration and/or temperature markedly reduced the contact time that was necessary. Increased chlorine dioxide concentration, temperature, and pH decreased the contact time that was required to produce 99% inactivation of the viruses. The amount of inactivation depended on the virus that was tested.

Mechanism of Action

There is little information concerning the mode of action by which chlorine dioxide inactivates bacteria and viruses. Ingols and Ridenour suggested that the bactericidal effectiveness of chlorine dioxide is due to its adsorption on the cell wall with subsequent penetration into the cell where it reacts with enzymes containing sulfhydryl groups.

Benarde et al. demonstrated that chlorine dioxide abruptly inhibited protein synthesis. The incorporation of ^{14}C -labeled amino acids into protein by whole cells stopped within a few seconds after the addition of chlorine dioxide. Subsequently, Olivieri reported a dose-response in the inhibition of protein synthesis in bacteria that had been treated with chlorine dioxide. The site of action was localized in the soluble portion (enzymes) of the cell extracts of treated cells without affecting the integrity of the ribosomes' function in protein synthesis.

Chlorine dioxide is an effective bactericide and virucide under the pH, temperature, and turbidity that are expected in the treatment of potable water. It should be noted that the U.S. Environmental Protection Agency has set an interim Maximum Contaminant Level of 1 mg/liter on chlorine dioxide because of the unresolved questions on its health effects.

A simple, selective, and sensitive test for chlorine dioxide should be developed to monitor residual concentration.

Chlorine dioxide is currently used at several plants. A review of plant records and field studies on the stability and effectiveness of chlorine dioxide in the distribution system should be undertaken.

Studies should be directed toward evaluating the inactivation of protozoan cysts.

More information should be obtained on the mode of action by which chlorine dioxide inactivates bacteria, viruses, and cysts.

Iodine

The use of iodine as a biocide has had a long history, primarily as an antiseptic for skin wounds and mucous surfaces of the body and, to a lesser degree, as a powerful sanitizing agent in hospitals and laboratories. The use of iodine as a disinfectant of drinking and swimming pool water has not been extensive, mainly because of the costs and problems that are involved in applying the dosage.

Aside from the emergency iodination of small volumes for field and emergency drinking water and limited experience with swimming pool disinfection, the only substantial experience with iodine disinfection of piped water system is that of Black et al. Two water systems serving three prisons in the state of Florida were disinfected satisfactorily with 1 mg/liter of iodine. A persistent residual was maintained throughout the distribution system despite a finished water pH of 8.0 to 9.5. No adverse effects on health were observed among those consuming the water.

Chemistry of Iodine in Water

Iodine is the only common halogen that is a solid at room temperature, and it possesses the highest atomic weight (126.91). Of the four common halogens, it is the least soluble in water, has the lowest standard oxidation potential for reduction to halide, and reacts least readily with organic compounds.

$$I_2 + H_2O \rightleftharpoons HOI + H^+ + I^-$$

Diatomic iodine (I_2) reacts with water to form hypoiodous acid (HOI) and iodide ion (I^-). The effect of pH on this reaction is shown in table.

Table: Hydrolysis of Iodine at 25 °C Showing the Percentage of Iodine Species at Different pH's.

	Percentage of Iodine Species by Concentration of Iodine								
	0.5 mg/liter			5.0 mg/liter			50.0 mg/liter		
pH	I_2	HOI	OI⁻	I_2	HOI	OI⁻	I_2	HOI	OI⁻
5	99	1	0	100	0	0	100	0	0
6	90	10	0	99	1	0	100	0	0
7	52	48	0	88	12	0	97	3	0
8	12	88	0.005	52	48	0	86	14	0

The distribution of chemical species of iodine given in table was taken from the calculations that were made by Chang from the equilibrium expression:

$$\frac{[HOI][H]^+[I]^-}{[I_2]} = Kh$$

The value of K_h, the hydrolysis constant, is given by Wyss and Strandskov as 3×10^{-12} at 25 °C. With iodine residuals at 0.5 mg/liter, which are expected in water systems, and a pH of 5, approximately 99% of the total iodine residual is present as iodine and only 1% as hypoiodous acid. At pH 7, the two forms are present in almost equal concentrations. At pH 8, only 12% is present as elemental iodine and 88% as hypoiodous acid, which can be converted to hypoiodite ion (OI^-).

$$HOI \rightleftharpoons H^+ + OI^-$$

$$\frac{[H^+][OI^-]}{[HOI]} = K_a = 4.5 \times 10^{-13}$$

$$\left[H^+\right] = K_a \frac{\left[HOI\right]}{\left[OI^-\right]}$$

$$\frac{\left[HOI-\right]}{\left[OI^-\right]} = \frac{H^+}{K_a}$$

The dissociation constant, K_a, of hypoiodous acid (at 20 °C) is 4.5×10^{-13}. Consequently, the dissociation of hypoiodous acid, which occurs at high pH's, is not important for practical purposes. However, as confirmed by the field studies in Florida, hypoiodous acid can form iodate ion by autooxidation at pH values above 9.

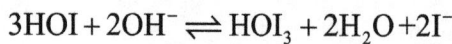

$$3HOI + 2OH^- \rightleftharpoons HOI_3 + 2H_2O + 2I^-$$

The iodate ion possesses no disinfecting ability.

Production and Application

Iodine may be added to a municipal water supply by several procedures. One method is to employ nonhazardous solvents and solubilizing agents such as ethyl alcohol $\left(C_2H_5OH\right)$ and potassium iodide (KI) to overcome the low concentration of aqueous iodine stock for solution feeders. Another method produces the required concentration of iodine by passing water through a bed of crystalline iodine (saturator). This has passing water through a bed of crystalline iodine (saturator). This has been used in many small, semipublic and private home water systems. Since the maximum concentration is limited by solubility to 200-300 mg/liter at the ambient temperatures that are expected for drinking water, some physical complications would accompany the introduction of this method into the large waterworks system in view of the large saturation beds required. The iodinated anion exchange resin bed and the vaporization technique are not sufficiently developed to be considered for use in public water supplies.

In certain circumstances, potassium iodide might be combined with an oxidation reaction to release iodine. The chemistry is simple, and the persistence of the iodine that is generated may be much better than chlorine. This method has been used in swimming pools and for dechlorination purposes where a chlorine residual may be exchanged for an iodine residual, or an iodine residual may be provided where ozone is the primary disinfectant.

Analytical Methods

Both amperometric titration and leuco crystal violet (LCV) colorimetric methods give acceptable results when used to measure free iodine in drinking water. Waters containing oxidized forms of manganese interfere with the LCV method. Also, when iodide ion exceeds 50 mg/liter and chloride ion exceeds 200 mg/liter, the amperometric method is preferred. Under unusual situations, where mixtures of chlorine, bromine, and ozone occur along with iodine, the problem of separation is difficult.

Biocidal Activity

Table shows the relative resistance of bacteria, viruses, and cysts to inactivation by iodine.

Table: Comparative Values from Confirmed Experiments on Disinfecting Water with Iodine at 23 °C to 30 °C, at pH 7.0.

Organism	Iodine mg/liter	Minutes for 99% Inactivation	c·t
Coliform bacteria	0.4	1	0.4
Poliovirus 1	20	1.5	30.0
f₂ Virus	10	3.0	30.0
Simian cysts	15	10.0	150.0

Efficacy against Bacteria

Chang and Morris, summarizing the development of a universal water disinfectant tablet for the military, concluded that iodine concentrations of 5 to 10 mg/liter were effective against bacterial pathogens. Iodine was less dependent than chlorine upon the pH, temperature, contact time, and secondary reactions with nitrogenous impurities in the water. As a cysticide, iodine was poor in water with a high pH. Consequently, the tablet that was formulated contained an acid buffer to lower the pH of the water. The tablet, called "Globaline," released 8 mg/liter of iodine, which is extremely high compared to the amount that is possibly needed for public water supplies. The tablet is not widely accepted, since color, taste, and odor problems are fairly common.

Chambers et al. investigated the effect of iodine concentration, pH, exposure time, and temperature on 13 enteric bacteria in a clean system. Their results are not completely quantitative in that the reported iodine concentration is that required to inactivate all test organisms plated out after 1, 2, and 5 min of contact. This was equivalent to approximately 99.9% activation with the procedure that they followed. There was definitely an observed pH effect. At 2 °C to 5 °C, some bacterial species required 3 to 4 times as much iodine for similar inactivation at pH 9.0 as was required at pH 7.5. For 5 min of exposure, the required dosage for ≥99.9% bacterial inactivation in high pH water at 20 °C-26 °C was always less than 1 mg/liter. A summary of their work with E. coli is given in Table. This is the best available information on iodine as a bactericide.

Table: Concentrations of Iodine and Contact Times Necessary for 99% Inactivation of Escherichia coli.

Iodine, mg/liter	Contact Time, min	c·t	pH	Temperature °C
1.3	1	1.3	6.5	2-5
0.9	2	1.8	6.5	
1.3	1	1.3	7.5	
0.7	2	1.4	7.5	
0.8	1	0.8	7.5	
0.6	2	1.2	7.5	
0.8	1	0.8	8.5	
0.9	2	1.8	8.5	
1.8		1.8	9.1	
1.2	2	2.4	9.1	
0.35		0.35	6.5	20-25
0.20	2	0.40	6.5	

0.45	1	0.20	7.5	
0.30	2	0.60	7.5	
0.45	1	0.45	8.5	
0.40	2	0.80	8.5	
0.45	1	0.20	9.1	
0.30	2	0.60	9.1	

Efficacy against Viruses

Studies on the efficacy of iodine on viruses have shown that viruses are more resistant to disinfection than are vegetative cells of bacteria. Krusé compared virus and bacterial inactivation. At an iodine dose of 10 mg/liter in 0.048 mM potassium iodide at room temperature and pH 7.0, E. coli was inactivated more rapidly than f2 virus but both were reduced by 4 logs in less than 1 min. However, when the potassium iodide concentration was raised to 0.5 M, bacterial inactivation was 4 logs in 1 min while virus inactivation was only 0.5 log in 1 hr.

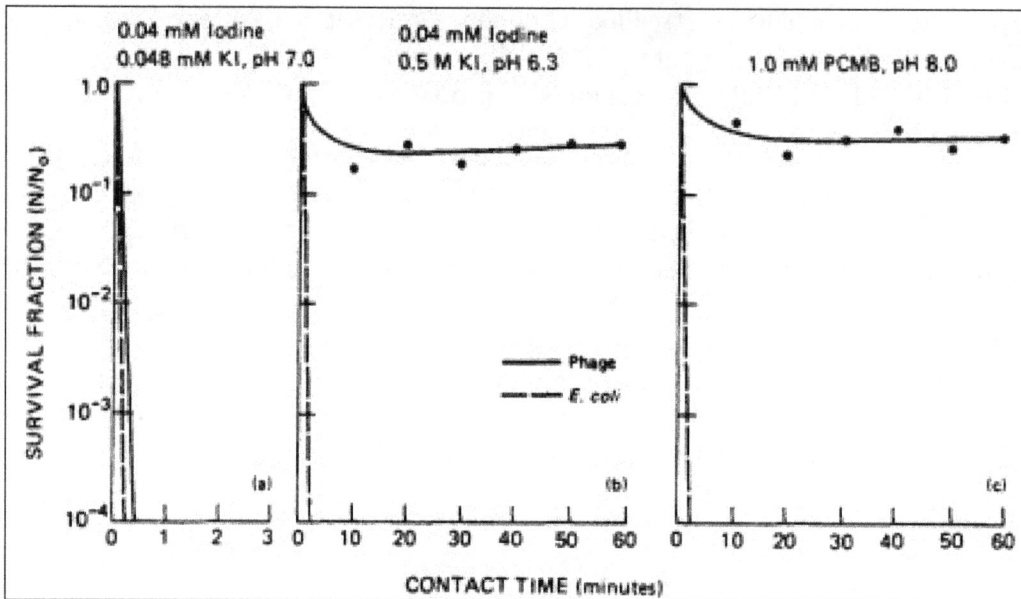

Survival of E. coli and f_2 bacteriophage reacted with 0.04 mM iodine containing (a) 0.048 mM potassium iodide (KI) at 37 °C and (b) 0.5 mM potassium iodide at 0 °C, and survival obtained with (c) 1 mM of a sulfhydryl reacting agent, p- chloromercuribenzoic acid ($ClHgC_6H_4COOH$).

Berg et al. measured the dynamics of survival for poliovirus, coxsackievirus, and echovirus that were iodinated at pH 6.0 and at temperatures of 5 °C, 15 °C, and 25 °C. The work of the Johns Hopkins group from 1962 to 1969 was primarily with the virus model f2, although some comparative poliovirus 1 work was done. Cramer et al. have shown that the mode of inactivation of these two viruses by iodine is similar. Data on the inactivation of virus by iodine are summarized in Table for both polio and f2. The results of inactivation studies by the various groups compare very favorably. At pH 6.0, iodine approaches the order of magnitude of virucidal activity of hypochlorous acid (HOCl). At the pH's likely to be maintained in the distribution system (pH 8.0), iodine is a vastly more effective virucide than combined chlorine and is not far removed from the activity range of free chlorine.

Table: Concentrations of Iodine and Contact Times Necessary for 99% Inactivation of Polio and f2 Viruses with Flash Mixing.

Test Microorganism	Iodine, mg/liter	Contact Time, min	c·t	pH	Temperature,°C
f_2 Virus	13	10	130	4.0	5
f_2 Virus	12	10	120	5.0	
f_2 Virus	7.5	10	75	6.0	
f_2 Virus	5	10	50	7.0	
f_2 Virus	3.3	10	33	8.0	
f_2 Virus	2.7	10	27	9.0	
f_2 Virus	2.5	10	25	10.0	
f_2 Virus	7.6	10	6	4.0	25-27
Poliovirus 1	30	3	90	4.0	
f_2 Virus	64	10	64	5.0	
f_2 Virus	4.0	10	40	6.0	
Poliovirus 1	1.25	39	49	6.0	
Poliovirus 1	6.35	9	57	6.0	
Poliovirus 1	12.7	5	63	6.0	
Poliovirus 1	38	1.6	60	6.0	
Poliovirus 1	30	2.0	60	6.0	
f_2 Virus	3.0	10	30	7.0	
Poliovirus 1	20	1.5	30	7.0	
f_2 Virus	2.5	10	25	8.0	
f_2 Virus	2.0	10	20	9.0	
f_2 Virus	1.5	10	10.0	15	
Poliovirus 1	30	0.5	15	10.0	

Efficacy against Parasites

Many studies on cyst inactivation have been reported, but there are discrepancies in their results due mainly to differences in the test systems that were used. The dose, pH, and temperature that were used in many of the studies are in doubt as are the different sources, cleaning methods for cysts, number of cysts used per test, and determination of viability. Stringer compared the resistance to iodine disinfection of Entamoeba histolytica cysts obtained from in-vitro culturing with those harvested from a human carrier and mixed amoebic cysts from monkeys. In water at room temperature at pH 6.5, 3 times the iodine dose was required in 10 min of exposure for 99% inactivation with wild (naturally formed) cysts compared to the cultured cysts. Large numbers of amoebic cysts from simian hosts were more readily available than cysts from human stools. They served as a reliable model for human E. histolytica.

The most extensive and reliable data on the cysticidal properties of iodine are found in the work of Chang and Stringer and their co-workers. The only quantitative comparative study of the cysticidal properties of chlorine, bromine, and iodine believed to be published is that of Stringer, who reported a 99.9% inactivation. The earlier studies of others involved "total kills" and are quite dependent on the cyst density that was used.

All investigators more or less agree that iodine is an excellent cysticide in low pH waters (< pH 4.0), suggesting that molecular iodine (I_2), rather than hypoiodous acid, is the active agent. Chang observed that when the titratable iodine dosages exceed 20 mg/liter in the presence of iodide ion, the triiodide ion (I_3^-) has a cysticidal efficiency that is equal to 1/11, 1/8, and 1/7 that of I_2 at 6 °C, 25 °C, and 35 °C. He further calculated relative cysticidal efficiency of hypoiodous acid to be one-third that of I_2 at 6 °C and one-half that of I_2 at 20 °C.

There may be the problem of ineffective iodine residual in the distribution system where cross-connection introduction of cyst contamination is a possibility. This is due to the practice (for corrosion control) of maintaining water in the distribution system at approximately pH 8.0. At the low halogen residual levels that are usually maintained, the most cysticidal form, I_2, would be less prevalent than hypoiodous acid. Water at 25 °C and pH 8.0 with a total iodine residual of 0.5, 1.0, and 2.0 mg/liter would contain only 0.06, 0.2, and 0.62 mg/liter of molecular iodine. Table shows the time in minutes required for 90% inactivation of simian cysts with residuals of bromine, chlorine, and iodine in water at pH 6.0, 7.0, and 8.0 and a temperature of 30 °C. At pH 8.0, iodine is inferior to free bromine and chlorine. However, in the presence of excess ammonia, with which the halogens could react, bromine appears to be more effective at pH 8.0 than do similar dosages of iodine or chlorine. Residual iodine in water at pH 8.0 has the ability to persist.

Table: Contact Times Necessary for Low Residual Bromine, Chlorine, and Iodine in Water at 30°C to Effect 99% Inactivation of Cysts from Simian Stools.

| | | Contact Time Required, min, by 10-min Halogen Residual Concentration, mg/liter | | | | | | | | |
| | | 0.5 mg/liter | | | 1.0 mg/liter | | | 2.0 mg/liter | | |
	pH	Br_2	Cl_2	I_2	Br_2	Cl_2	I_2	Br_2	Cl_2	I_2
Buffered water	6.0	10	10	20	4	4	10	3	3	5
	7.0	12	14	40	8	12	20	4	5	7
	8.0	15	15	ND	10	15	80	5	10	20
Buffered water in presence of excess ammonia	6.0	10	65	20	8	35	10	4	22	5
	7.0	30	120	40	10	55	20	7	35	7
	8.0	35	ND	ND	13	80	80	9	50	20

The kinetics of iodine as a cysticide in water at 20 °C to 30 °C have been assembled and compared. The cultured E. histolytica cyst data of Chang and co-workers and the results of Stringer et al. present a consistent picture. For cultured cysts in water at pH 5 the c · t values for 99% inactivation calculated by Chang were 200 at 3 °C, 130 at 10 °C, and 65 at 23 °C.

These c · t values from Stringer et al. were approximately one-half the values they obtained with E. histolytica cysts from human stools and for mixed cysts from simian stools.

Mechanism of Action

The failure of a strong commercial iodine disinfectant to inactivate poliovirus led to interest in the responsible mechanism, especially the role of pH and iodide ion. Berg et al. claimed that iodine inactivation of coxsackievirus resulted from bio-molecular reaction with a single iodine molecule and that clumping of virions played a role in resistance to disinfection. Fraenkel-Conrat pointed out inconsistencies in the literature regarding the virucidal properties of iodine. Hsu and Hsu *et al.* clarified the mechanism of iodine on cells and virus. Hsu extracted fully active transforming DNA from iodine-inactivated *Haemophilus parainfluenzae* cells and just as much infectious RNA from f2 bacterial virus that had been inactivated (5 logs) by iodine as from noniodinated controls. Iodine inactivation, unlike the action of chlorine, appears to be attributable to a reaction with vital amino acids in proteins. Further experiments were conducted to determine whether sulfhydryl, tryptophanyl, histidyl, or tyrosyl groups were involved. With bacterial cells, there was a striking similarity between the kinetics of inactivation with iodine and the application of p-chloromercuribenzoic acid, a sulfhydryl reacting agent. While the active chemical species of iodine is not known, Hughes, Allen and Keefer, Bell and Gelles, and Hsu hypothesized that the hydrated cationic iodine species $\left(H_2OI^+\right)$ attacks the base, first against the sulfhydroxyl groups, and is not materially affected by the presence of iodide ion as was tyrosine. When sulfhydroxyl groups are the site of inactivation, a low pH should favor the reaction. With the viruses, the sulfhydroxyl group is not involved. Evidence of tyrosine's involvement came from parallel experiments showing similar patterns of inactivation curves between iodination of f2 virus and $_L$ tyrosine. Li had shown that the presence of iodide ion and low pH iodine $\left(I_2\right)$ inhibited the iodination of tyrosine. Both poliovirus and f2 virus inactivation with iodine was inhibited by the iodide ion. At pH 4.0, both polio and f2 viruses survived iodination whereas at pH 10 inactivation was complete. Although iodine decomposes rapidly to iodate $\left(IO_3^-\right)$ and iodide at pH 10, flash mixing of iodine (yielding hypoiodous acid) overcomes this difficulty effectively, and the virus inactivation is complete in less than 1 min. Therefore, there is evidence that iodine action, with little or no iodide present, is effected by the modification of protein without destroying DNA or RNA. The mode of action of iodine in cyst penetration and inactivation has not been studied.

Iodine has many features that are comparable to free chlorine and bromine as a water disinfectant, but iodamines are not formed. Free iodine is an effective bactericide over a relatively wide range of pH. Field studies on small public water systems have shown that low levels of 0.5 to 1 mg/liter of free iodine can be maintained in distribution systems and that the magnitude of residual is sufficient to produce safe drinking water with no adverse effects on human health. Like other halogens, the effectiveness of iodine against bacteria and cysts is significantly reduced by high pH, but unlike bromine and chlorine it is much more effective against viruses because of the enhanced iodination of tyrosine. Currently, its use is restricted primarily to emergency disinfection of field water supplies because of its high cost and because it is difficult to apply to large systems. The possible adverse health effects of increased iodide intake for susceptible individuals in the population must also be considered.

Studies should be conducted to determine the consequences for human health of the long-term consumption of iodine in drinking water with special regard for more susceptible subgroups of the population.

Bromine

Chemistry of Bromine in Water

Bromine was first applied to water as a disinfectant in the form of liquid bromine $\left(Br_2\right)$, but it can also be added as bromine chloride gas (BrCl) or from a solid brominated ion exchange resin.

Oxidation of bromide (Br^-) to bromine can also be accomplished either chemically or electro-chemically. Oxidation with aqueous chlorine gives either bromine or hypobromous acid (HOBr), depending on the ratio of chlorine to bromide. Both bromine and bromine chloride hydrolyze to hypobromous acid:

$$Br_2 + H_2O \rightarrow HOBr + H^+ + Br^-$$

$$BrCl + H_2O \rightarrow HOBr + H^+ + Cl^-$$

Molecular bromine exists in water at moderately acid pH and high bromide concentrations since the equilibrium constant of Reaction 21 is 5.8×10^{-9} at 25 °C. Like chlorine, bromine chloride has a much higher hydrolysis constant than this, so it does not exist as the molecular form in appreciable concentrations under conditions of water treatment. The ratio of molecular bromine to hypobromous acid depends on both pH and bromide concentration. From the equilibrium expression for Reaction $Br_2 + H_2O \rightarrow HOBr + H^+ + Br^-$:

$$\log \frac{(Br_2)}{(HOBr)} = \log(Br^-) - pH + 8.24$$

Thus, for a solution containing 10 mg/liter bromide and a pH of 6.3, 1% of the bromine is Br_2, while at lower bromide concentrations or higher pH aqueous bromine occurs almost entirely as hypobromous acid, which is a very weak acid with a dissociation constant of 2×10^{-9} at 25 °C.

$$HOBr \rightleftharpoons H^+ + OBr^-$$

The hypobromite ion (OBr^-) becomes the major form of bromine above pH 8.7 at 25 °C. Lower temperature decreases both of the above equilibrium values, thereby increasing the pH range where hypobromous acid is the major chemical form of bromine in water.

Bromine and bromine chloride also react with basic nitrogen compounds to form combined bromine or bromamines:

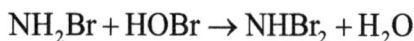

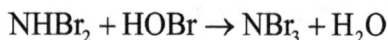

$$HOBr + NH_3 \rightarrow NH_2Br + H_2O$$

$$NH_2Br + HOBr \rightarrow NHBr_2 + H_2O$$

$$NHBr_2 + HOBr \rightarrow NBr_3 + H_2O$$

The observed breakpoint for ammonia (NH_3) solutions that have been treated with bromine is similar to that seen with chlorine. For bromine, this point corresponds to 17 mg/liter bromine for 1 mg/liter of ammonia nitrogen $(NH_3 - N)$. At this point, a minimum of bromamine stability occurs as ammonia nitrogen is oxidized to nitrogen gas:

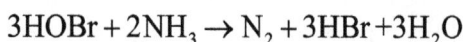

$$3HOBr + 2NH_3 \rightarrow N_2 + 3HBr + 3H_2O$$

At bromine-to-ammonia ratios higher than this and in the acid pH range, nitrogen tribromide

(NBr_3) is stable. It is the most abundant bromamine in such aqueous solutions. At lower bromine concentrations, dibromamine $(NHBr_2)$ can be present, but it is quite unstable. Only at alkaline pH values and very high ammonia concentrations such as those found in wastewater are significant quantities of monobromamine (NH_2Br) formed. Organic bromamines are also formed, but there is little information on their forms or stability.

Production and Application

Bromine is produced by oxidation of bromide-rich brines (that contain between 0.05% and 0.6% bromide) with chlorine. Bromine is then stripped with steam or air and is collected as liquid Br_2.

Bromine chloride is produced by mixing equal molar quantities of pure bromine and chlorine. It condenses to liquid bromine chloride below 5 °C at I atm pressure or above 30 psig at 25 °C. In the liquid phase, more than 80% of the liquid is bromine chloride, and the remainder is Br_2 and Cl_2. In the gas phase, 40% of the bromine chloride dissociates to Br_2 and Cl_2 at 25 °C.

Bromine has been applied to water as liquid Br_2. The difficulties that are encountered when handling the liquid and its corrosive nature, especially when wet, have encouraged the use of bromine chloride. Liquid bromine chloride is removed from cylinders under moderate pressure. It is then vaporized, and the gas is metered in equipment that is similar to that used for chlorine. Like chlorine, bromine chloride is shipped as the dry liquid in steel containers. Gas feeders must be made of Teflon, Kynar, or Viton plastics because bromine chloride is more reactive than chlorine with polyvinylchloride plastics.

Analytical Methods

Bromine concentrations can be measured iodometrically by procedures that are identical to those used to measure total chlorine residuals. None of these procedures is capable of distinguishing free bromine $(Br_2, HOBr, OBr^-)$ from combined bromine (bromamines) or other oxidants that are capable of reacting with iodide under the slightly acid conditions used in these procedures. However, bromate does not interfere except at low pH. UV spectroscopy can be used to measure the bromamines selectively in the presence of one another and without interference from free bromine because none of the free bromine forms except Br_3^- has strong UV absorptivities.

Biocidal Activity

Since residual bromine was rarely measured in the studies cited here, $c \cdot t$ (concentration, mg/liter, times contact time, min) products have generally been calculated using the dosage of bromine added to the system.

Efficacy against Bacteria

The efficacy of bromine inactivation of bacteria has been summarized by Farkas-Hinsley. Vegetative cells are readily inactivated by bromine, but reports often leave the type of bromine compound and its residual concentration as uncertain. Consequently, it is difficult to make quantitative comparisons with other disinfectants. Tanner and Pitner reported that the concentrations of hypobromous acid that are required to give "complete kill" in 30 min are 0.15 mg/liter for Escherichia coli and 0.6 mg/liter for Salmonella typhi. Spores of Bacillus subtilis required more than 150 mg/liter of bromine. Krusé et al. found that 4 mg of bromine per liter as hypobromous acid gave 5 logs of disinfection of E. coli in 10 min at 0 °C. At pH 4.5, where significant Br_2 was present, 2 mg/liter bromine at 0 °C gave 4 logs of E. coli disinfection in 3 min. However, Krusé also reported that 0.1 M bromide decreased E. coli disinfection to 2 logs of inactivation from 4.5 logs at 0.001 M bromide, using 4 mg/liter bromine at pH 7.5 and 0 °C in each case. This is in conflict with observations made at low pH, since high bromide should also produce bromine. The data at pH 4.5 showed 4 logs inactivation with a c · t of 6 compared to 5 logs inactivation from a c · t of 40 for hypobromous acid. Even these data, the best available, are based on the concentrations of bromine added to the solution. It is difficult to compare the various studies on disinfection by bromine, since the chemical species of bromine present in most instances have not been measured or reported.

The effect of the formation and decomposition of the bromamines on the efficacy of bacterial disinfection was first discussed by Johannesson. He reported that 0.28 mg/liter of monobromamine expressed as bromine resulted in 99% inactivation in less than 1 min, while the same concentration of N-bromodimethylamine $\left[N(CH_3)_2 Br \right]$ required 12 min for 99% inactivation of E. coli. The measurements of the halogen remaining after the experiments showed very little loss.

The efficacy of bromine inactivation of spores of Bacillus metiens and B. subtilis has been studied by Marks and Strandskov and Wyss and Stockton. Both pairs of investigators found that the activity was markedly pH-dependent at low pH values where molecular bromine predominates. The activity increased rapidly from pH 4 to pH 3, the lowest that was tested. The c · t required for 99% inactivation was 20-25 at 25 °C and pH 3.0. At pH 6 to 8, the c · t was 225 to 375. At pH 9, where the hypobromite ion starts to predominate, the c · t increased to more than 500.

Marks and Strandskov also determined that at pH 7 and 25 °C monobromamine was one-half as effective as hypobromous acid against B. metiens spores. N-bromosuccinimide was 1/17, and N-bromopiperidine was 1/800 as effective. Wyss and Stockton showed the effect of changing chemical form and concentration by increasing the concentration of ammonia that was added to a 20 mg/liter bromine solution at pH 7 containing B. subtilis spores. Table shows the measured residual and the time required for 99% inactivation. Before the ammonia is added to the solution, hypobromous acid is the major chemical form. When the ammonia nitrogen reaches 1 mg/liter, nitrogen tribromide predominates. At 2 mg/liter ammonia nitrogen, the breakpoint occurs with rapid loss of combined bromine residual. At 10 and 30 mg/liter ammonia nitrogen, dibromamine is present and is less stable and effective than hypobromous acid. At 1,000 mg/liter ammonia nitrogen, monobromamine has a c · t value for 99% inactivation that is as effective as nitrogen tribromide and hypobromous acid.

Table: The Effect of Ammonia on Bromine Concentrations Required for 99% Inactivation of Bacillus subtilis Spores at 20 mg/liter, pH 7, 25 °C.

Ammonia, mg NH_3-N/liter	10-min Residual, mg/liter	Contact Time, min	c·t
0	20	14	280
1	16	19	304
2	2	> 100	> 200

10	8	85	680
30	9	70	630
100	14	18	252
1,000	19	14	266

The effect of temperature on the efficacy of hypobromous acid inactivation of B. subtilis spores is shown in table. The activity increases approximately 2.3 times for each 10 °C rise in temperature. After measuring washed vegetative cells of B. subtilis, Wyss and Stockton also found them to be 500 times less resistant. They reported no temperature data for their vegetative cell studies.

Table: The Effect of Temperature on Bromine Concentrations Required for 99% Inactivation of Bacillus subtilis Spores with 25 mg/liter Bromine at pH 7.0.

Temperature, °C	Contact Time, min	c·t
35	4	100
25	10	250
20	16	400
15	21	525
10	37	925
5	54	1,350

Efficacy against Virus

Krusé et al. reported that 4 mg/liter of bromine as hypobromous acid at pH 7 and 0 °C gave 3.7 logs of inactivation of f2 E. coli phage virus in 10 min. At higher bromide concentrations and lower pH, where Br_2 becomes the principal form of bromine, the rates of inactivation increased. A c · t of 40 was required for 3.7 logs inactivation at pH 7.0 and 0°C compared to a c · t of 1 for 4.5 logs inactivation at 0 °C and pH 4.8. At pH 7.5 and with a concentration of 0.1 M bromide at 0 °C, a c · t of 12 yielded 5.5 logs inactivation. In these studies the residual concentrations of bromine were probably not significantly different from those in solutions that did not contain added salts. In the presence of added bromide, amines were difficult to quantify. After 2 hr, Krusé and his colleagues found no viable virus in a solution of 4 mg/liter bromine, starting from an initial titer of 1.7×10^9 plaque-forming units (PFU)/ml at 0 °C and pH 7.5, when excess ammonia was present. However, the addition of bromine to phage solutions that contained excess methylamine (CH_3NH_2) and glycine (H_2NCH_2COOH) essentially stopped disinfection under these conditions. This may be due either to the formation of the N-bromo compounds or to the reduction of bromine.

Taylor and Johnson used Fx174 E. coli phage and measured concentrations of the disinfecting species in constant residual solutions. They reported that 0.32 mg/liter hypobromous acid required only 1.1 min or a c · t of 0.35 to inactivate 99% of this phage at 0 °C. Compared to hypobromous acid, molecular bromine was approximately 3 times as fast, requiring Br_2 at c · t of 0.1. Nitrogen tribromide was less potent, requiring a c · t of 1.0 for 99% inactivation of this phage at 0 °C.

The effect of temperature on Fx174 inactivation was measured for 0.16 mg/liter of hypobromous acid expressed as bromine at pH 7. The Arrhenius plot of In k against 1/T (where T is temperature) from 273 °K to 303 °K gave a slope equivalent to 37 kcal/mol or an increase in rate of inactivation of 1.9 times for a rise in temperature from 15 °C to 25 °C.

Sharp's group has made careful studies with measured, controlled residual concentrations. They demonstrated that even under these controlled chemical conditions the degree of aggregation or clumping had a marked effect on the apparent, observed inactivation rates with reovirus and type 1 poliovirus. Reoviruses, as single particles, required only 1 s for 3 logs of inactivation at pH 7 and 2 °C with hypobromous acid at 0.46 mg/liter bromine. Aggregated samples gave much slower rates, especially at high levels of inactivation. At 3 logs of inactivation, the time required doubled for the same concentration conditions. Poliovirus 1 was also rapidly inactivated by hypobromous acid when single virus particles were studied. At 2 °C and pH 7, only 3.5 mg/liter bromine or a c · t of 0.2 was required for 2 logs of inactivation. However, the rate of inactivation did not increase linearly with concentration at this temperature. Higher concentrations were much less efficient than longer exposures. At 1 mg/liter hypobromous acid, 7 s were required for 1 log of inactivation or a c · t of 0.23 for 2 logs of inactivation at 2 °C and pH 7 with hypobromous acid. The inactivation rate increased only slightly with temperature at this concentration. At 10 °C a c · t of 0.21 for 2 logs of inactivation and at 20 °C a c · t of 0.06 was required for 99% inactivation with hypobromous acid.

The inactivation of single poliovirus particles in buffered, distilled water and constant residual concentrations for the other major bromine chemical forms have also been studied by Floyd et al. Table gives the calculated c · t required to yield 99% inactivation for these different forms near 1 mg/liter expressed as bromine. The values in Table depend on concentration except for Br_2, for which time and concentration were inversely related as normally assumed from the Watson-Chick relationship. Floyd et al. also demonstrated that dibromamine, nitrogen tribromide, and hypobromous acid were less efficient at higher concentrations while the hypobromite ion became more effective as the concentration increased. The rate for the hypobromite ion is rapid compared to the other bromine compounds. This is interesting in light of the fact the hypochlorite ion (OCl^-) is generally considered to be a much poorer disinfectant than hypochlorous acid (HOCl).

Table: Exposure (c·t) to Various Bromine Compounds Required for 99% Inactivation of Poliovirus 1, Mahoney.

Chemical Form	c·t	Temperature, °C	pH
Dibromamine ($NHBr_2$)	1.2	4	7.0
Nitrogen tribromide (NBr_3)	0.19	5	7.0
Bromine (Br_2)	0.03	4	5.0
Hypobromite ion (OBr^-)	0.01	2	10.0
Hypobromous acid (HOBr)	0.24	2	7.0
Hypobromous acid	0.21	10	7.0
Hypobromous acid	0.06	20	7.0

Efficacy against Parasites

There appears to be no information regarding the effectiveness of bromine as a disinfectant against eggs or larvae of helminth parasites in water.

With protozoa that are important to public health, attention appears to have been devoted primarily to the efficacy of bromine in destroying cysts of Entamoeba histolytica. Most relevant information has been provided by Stringer et al. as a result of their studies on the comparative cysticidal efficacies of various halogens. They used bromine stock solutions that were prepared by bubbling nitrogen through bottles of elemental bromine. Concentrations were determined by a colorimetric bromocresol purple test. Cyst survival was evaluated by the excystation method that these investigators had developed.

In dose-response experiments in distilled water for exposures of 10 min and at pH levels from pH 4 to pH 10, bromine was found by Stringer et al. to be the most effective and fastest acting of the halogens tested over the widest pH range. At pH 4, 99.9% cyst mortality was obtained with 1.5 mg/liter of free bromine residual; whereas 2 mg/liter of chlorine and 5 mg/liter of iodine were required to attain the same mortality. Furthermore, increases in pH seemed to have less effect on the cysticidal efficacy of bromine as compared with the other halogens. At pH 10, 99.9% mortality was obtained with residuals of 4 mg/liter of bromine, 12 mg/liter of chlorine, and 20 mg/liter of iodine.

In studies more nearly simulating usual water treatment procedures, "flash mixing" of halogens at a dosage of 2 mg/liter with the cyst suspensions was used. In buffered distilled water, bromine again proved to be the most effective: 99.9% inactivation was obtained at pH 6 and pH 8 after 15 min of contact. Longer exposures at the lower pH were required for the other halogens in order to provide even less inactivation than this.

An interesting aspect of the study of Stringer et al. was that ammonia bromamines were nearly as cysticidal as free bromine except in waters of high pH.

Thus, under conditions likely to be found during the treatment of natural water supplies, free and combined bromine appear to be a practical, effective cystide, at least as far as cysts of E. histolytica are concerned.

Mechanisms of Action

The pattern of bromine disinfection appears to be similar to that of chlorine. After comparing the activity of chlorine, bromine, and iodine against spores, Marks and Strandskov noted that Br_2 was 9 times more effective than hypobromous acid and that the hypobromite ion and tribromide ion (Br_3^-) had very low activity. They noted that a high degree of polarity contributed to the inactivity of the ionic forms and the reduced activity of the hypohalous acid compared to the free molecular halogen. They also found that "the killing rates of bacterial spores for the hypohalous acids and probably for the molecular halogens decrease in the following order: chlorine, bromine, iodine."

Many workers attribute this decrease to the greater oxidation potential of the halogens with lower molecular weight. However, the polarity and perhaps the sizes of the halogens may be important in getting the disinfectants to the vital site.

Olivieri et al. demonstrated that bromine was effective in inactivating both naked viral RNA and intact virus. The primary site of inactivation of f2 phage more likely involves the reaction of bromine with the protein coat of the virus, because inactivation of RNA that was prepared from bromine-treated virus lagged significantly behind the inactivation of intact virus. The mechanism of inactivation with bromine, as with the other halogens, involves moving the disinfectant to the vital site, mass transport, and the reactivity of the bromine with that site, oxidation. The effectiveness of bromamines as disinfectants may be explained by their relatively low polarity and high reactivity. The effectiveness of the hypobromite ion as a virucide may be due to the fact the high pH›s of the solutions make the virus more sensitive to the ion. Or, it could be attributed to Olivieri›s observation that bromine acts primarily on the protein coat of viruses.

Laboratory tests show that bromine is an effective bactericide and virucide. It is more effective than chlorine in the presence of ammonia.

As a cysticide, it is highly active. Bromine is active over a relatively wide range of pH, and it retains some of its effectiveness as hypobromous acid to above pH 9.

The major disadvantage of bromine as a disinfectant is its reactivity with ammonia or other amines that may seriously limit its effectiveness under conditions that are encountered in the treatment of drinking water. Data on the effectiveness of bromine against bacteria are complicated by this reactivity and the lack of characterization of the residual species in disinfection studies.

Further research is required to quantitate the effectiveness of the various species of bromine and the bromamines, both organic and inorganic, against bacteria, viruses, and cysts, particularly with bacteria.

The reactivity of bromine and the dosages that are required to maintain effective residual in natural water systems should be investigated.

The effectiveness of various technologies for application of bromine or bromine chloride to intended drinking water should be evaluated.

Ferrate

Ferrates are salts of ferric acid (H_2FeO_4) in which iron is hexavalent. Fremy first synthesized potassium ferrate (K_2FeO_4) in the mid-nineteenth century. Since then, a wide variety of metallic salts have been prepared. However, only a few of the preparations yield ferrates of sufficient purity and stability for use in the treatment of water. Ferrates are strong oxidizing agents that have a redox potential of - 2.2 V or 0.7 V in acid and base, respectively.

The Chemistry of Ferrate in Water

Aqueous solutions of potassium ferrate are unstable and decompose to yield oxygen (O_2), hydroxide (OH-), and insoluble hydrous iron oxide [FeO(OH)]:

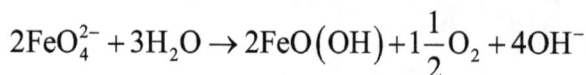

$$2FeO_4^{2-} + 3H_2O \rightarrow 2FeO(OH) + 1\frac{1}{2}O_2 + 4OH^-$$

The hydrous iron oxide is a coagulant that is commonly used in water treatment. Initial ferrate concentration, pH, temperature, and the surface character of the resulting hydrous iron oxide affect the rate of decomposition. Ferrate solutions are most stable in strong base (> 3 M or at pH

10 to 11). Schreyer and Ockerman reported that dilute aqueous solutions of ferrate are more stable than concentrated solutions. The presence of other inorganic ions also affects ferrate stability in aqueous solutions.

Ferrate reacts rapidly with reducing agents in solutions. It will also oxidize ammonia (NH_3). The rates for this reaction increase with pH (optimum pH range 9.5-11.2), molar ratio of ferrate to ammonia, and temperature. The oxidation of ammonia by ferrate below pH 9 is markedly slower than that observed for chlorine. Some information is available on the reaction of ferrates with various organic compounds.

Preparation

Concentrated solutions of sodium ferrate (Na_2FeO_4) may be prepared electrochemically from the more common iron forms. Scrap iron can be converted to ferrate iron with a 40% efficiency. Potassium ferrate may be prepared by wet oxidation of Fe(III) with potassium permanganate $(KMnO_4)$. Subsequent recrystallization yields a crystalline solid with greater than 90% purity. Ferrate salts are not commercially available. Consequently, sufficient quantities for pilot or full-scale testing would have to be especially prepared.

Analytical Methods and Residuals

Aqueous ferrate solutions have a characteristic violet color that is similar to permanganate and a wavelength maximum at 505 nm in the visible portion of the electromagnetic spectrum. The molar extinction coefficient at 505 nm in 10^{-4} M sodium hydroxide (NaOH) was $1,070 \pm 30$ 1 $M^{-7}cm^{-1}$. Since ferrates are unstable, residual cannot be maintained.

Biocidal Activity

Efficacy against Bacteria

Gilbert et al. reported the inactivation of a pure culture of Escherichia coli at pH 8.0. Increased ferrate concentrations from 1.2 to 6.0 mg/liter yielded increased rates of inactivation of E. coli. They observed 99% inactivation at 6.0 mg/liter, pH 8.0, and 27 °C in approximately 8.5 min. The rates of inactivation of E. coli with ferrate appear to be of the same order of magnitude as monochloramine (NH_2Cl). Waite extended Gilbert's disinfection studies to include enteric pathogens and Gram-positive bacteria and evaluated the effects of pH and temperature. The rate of E. coli inactivation by ferrate increases as pH decreases from pH 8.0 to pH 6.0. At the lower pH value, the rate of ferrate decomposition is increased, and little inactivation was observed after 5 min. Several inconsistencies at low ferrate concentrations (0.12 mg/liter) were observed, but low-level inactivations (< 50% inactivation) are difficult to interpret. At lower temperatures, the biocidal activities of ferrate appear to increase as temperature increases. At higher temperatures, the ferrate decomposition becomes an important factor. Salmonella typhimurium and Shigella flexneri were inactivated in a manner similar to that of E. coli. However, considerably higher dosages of ferrate (> 12 mg/liter) were necessary for inactivation of Streptococcus faecalis. Ferrate concentrations of 12 mg/liter and 60 mg/liter required 5 min and 15 min, respectively, for 99% inactivation at pH 8.0. Other Gram-positive bacteria tested (Bacillus cereus, Streptococcus bovis, and Staphylococcus aureus) were noticeably more resistant to ferrates than were the enteric bacteria. Higher doses of ferrate (15-20 mg/liter) were also necessary to inactivate 99% to 99.9% of the bacteria in the presence of organic material in wastewater.

Efficacy against Viruses

Waite reported that the inactivation of the RNA bacterial f2 virus with ferrate appeared to be more dependent on pH than was bacterial inactivation. Considerably more viral inactivation was observed at pH 6.0 than at pH 8.0. At 1.2 mg/liter ferrate, just under 4 min were required for 99% inactivation, while for similar conditions 13 min were necessary at pH 8.0.

The disinfection efficacy of ferrate is summarized in Table. The ferrate dose is given in the table, since ferrate residuals were not reported. Ferrate decomposes rapidly in aqueous solution, and the dose represents an overestimate of ferrate concentration. Nevertheless, the $c \cdot t$ products (concentration, mg/liter, times contact time, min) indicate the difference in efficacy relative to the test microorganism and provide a crude idea of the $c \cdot t$ product for ferrate.

Table: Concentrations of Ferrate $\left(FeO_4 2^-\right)$ and Contact Times Necessary for 99% Inactivation of Escherichia coli, Streptococcus faecalis, and f_2 Virus.

Test Microorganism	Ferrate Dosage, mg/liter	Contact Time, min	$c \cdot t$	pH	Temperature, °C
Escherichia coli	1.2	4.6	5.52	7.0	20
	6.0	2.5	15.0		
Streptococcus faecalis	1.2	365	438	7.0	20
	6.0	4.0	24		
f_2 Virus	1.2	3.7	4.4	6.0	20
	6.0	1.5	9		

While not as biocidal as the free halogens, chlorine dioxide, or ozone, ferrate appears to be similar to or slightly better than the chloramines as a bactericide and more active as a virucide than the chloramines. The combined application of ferrate as a disinfectant and a coagulant makes it an attractive alternate biocide.

High pH Conditions

High pH values are obtained when drinking water is softened by the commonly used precipitation method, which removes calcium $\left(Ca^{2+}\right)$ and magnesium $\left(Mg^{2+}\right)$ ions that cause water hardness.

In such cases, calcium hydroxide $\left[Ca\left(OH\right)_2\right]$ is the usual source of hydroxide that is used to raise the pH. In this process, pH values as high as 10.5 are reached and maintained up to 6 hr. Low pH conditions are not used in water treatment processes. Therefore, while low pH's are lethal to most microorganisms, they are of less interest in this report than high pH conditions, which might have potential for disinfection through modification of current practices.

Attainment of Elevated pH

High pH values in water are obtained by using either calcium hydroxide or sodium hydroxide (NaOH). Sodium hydroxide is a by-product of the production of chlorine by the electrolysis method.

Calcium hydroxide is prepared by reacting calcium oxide (CaO) with water. The calcium oxide is produced by heating calcium carbonate $(CaCO_3)$ to drive off the carbon dioxide (CO_2). The calcium oxide is normally prepared off the site of the water treatment plant. However, it is often slaked on site at all but the smallest plants, which purchase calcium hydroxide and use it directly. A few, very large water treatment plants produce their own calcium hydroxide by calcining their water-softening sludge.

Hydroxide (OH^-) is added to water as a water slurry containing calcium hydroxide or as a water solution of sodium hydroxide. In the usual water-softening technique, the water is flocculated for 10-30 min to promote the precipitation reactions. The precipitates are then removed by sedimentation with retention times from 1 to 6 hr.

Disinfection with high pH is readily accomplished, but the required pH values are so high that the pH must be reduced before the water is consumed. In a full-scale water treatment plant, this is accomplished in the recarbonation stage, which is a normal component of the water-softening process. However, for small institutions or private residences the pH reduction requirement would mitigate against the use of this disinfection technique.

Analytical Methods and Process Control

The pH can be reliably measured with the glass electrode at pH values of approximately 7 and at room temperatures. Special care must be taken to obtain accurate values at other temperatures and at the elevated pH values that are necessary for disinfection with this process.

Residual

The hydroxide that is used to obtain the necessary pH values enters into few side reactions in natural water after the initial formation of metallic hydroxides and the reactions with carbon dioxide. The amounts of hydroxide for these reactions can be readily computed after the water has been chemically analyzed. There is a slow additional loss of hydroxide from the water by reaction with carbon dioxide entering from the atmosphere. However, this is not a serious problem.

Biocidal Activity

The microbial inactivation under consideration in this chapter is that caused solely by high pH. When the pH of water is raised, the opportunity exists for precipitation of many compounds associated with carbonates, oxides, and hydroxides, among many others. Even in distilled water, calcium carbonate may precipitate if calcium hydroxide is used to raise the pH and if sufficient carbon dioxide enters the water during the experiment. These precipitates provide opportunities for adsorption and coagulation of microorganisms and will cause an increased removal over that obtained from the pH effect alone without the precipitate. The research cited below, unless otherwise noted, was conducted with procedures that obviated this problem or that made its effect insignificant.

Efficacy against Bacteria

Wattie and Chambers determined the times required to obtain 100% inactivation of several bacterial species at initial concentrations of approximately 1,500 organisms/ml in 20 °C to 25 °C dechlorinated tapwater, using calcium hydroxide for pH adjustment. The inactivation was significantly less at each pH for each organism between 0 °C to 1 C. At a pH of 11.5, the time for 100% inactivation of Escherichia coli was increased from 210 min between 20 °C to 25 °C to 355 min between 0°C to 1°C. The time for 100% inactivation of Salmonella typhi was increased from 75 min to 270 min for the same temperature change. No mention was made of difficulties from the formation of precipitates.

Table: Contact Times Necessary for High pH Conditions to Effect 100% Inactivation of Initial Concentrations of 1,500 Organisms/ml.

pH	Contact Time Required for 100% Inactivation by High pH Conditions, min				
	Escherichia coli	Enterobacter aerogenes	Pseudomonas aeruginosa	Salmonella typhi	Shigella dysenteriae
9.01-9.5	> 540	—	—	> 540	—
9.51-10.0	> 600	> 600	420	> 540	> 300
10.01-10.5	> 600	> 600	300	> 540	> 300
10.51-11.0	600	> 540	240	240	180
11.01-11.5	300	> 600	120	120	75

Riehl et al. reported E. coli inactivations of 95% in 8 hr at pH 10.5 and 5 °C, 100% in 2 hr at pH 10.5 and 15 °C, and 100% in approximately 30 min at pH 10.5 and 25 °C in distilled water. Calcium hydroxide was used for pH adjustment. Inactivations of 50% and 55% in 10 hr were noted for Salmonella montevideo and S. typhi, respectively, at 2 °C and pH 10.6 in distilled water. Initial concentrations of organisms were approximately 1,000/ml. They also observed that higher temperatures gave more inactivation at the same pH and contact time and that the composition of the water did not markedly influence the bacterial survival.

Berg and Berman determined the inactivation of E. coli at 25 °C in the dilution water medium that is described in the 11th edition of Standard Methods. Sodium hydroxide was used to adjust the pH.

Table: Inactivation of Escherichia coli by High pH Conditions at 25 °C in a Dilution Medium.

pH	Contact time, min	Inactivation, %
11.55	50	~10
11.70	50	~90
11.81	35	96
12.01	10	99.8
12.04	6	97

Efficacy against Viruses

Wentworth et al. reported that little or no poliovius 1 was inactivated in distilled water at pH 11.2 at room temperature in 90 min when calcium hydroxide or sodium hydroxide was used to increase the pH. For poliovirus 1 in distilled water without added salts, Sproul et al. showed that no inactivation occurred in 90 min at pH values of 10.5 or less when calcium hydroxide and sodium hydroxide furnished the hydroxide. Their work was done at 21 °C to 22 °C. Sproul obtained poliovirus 1 inactivations in 30 min of 7% at pH 11.5, 59% at pH 11.9 in 30 min, 94% at pH 12.1 in 20 min, and 99.83% at pH 12.5 in 5 min. He used distilled water with 100 mg/liter of sodium chloride (NaCl) at 22 °C to 23 °C.

Berg and Berman showed the inactivation of echovirus 7 AGKP8A1 at 25 °C in a dilution water medium. They used sodium hydroxide to adjust pH. From his work with the f2 bacteriophage, Donovan suggested that a calcium-virus complex was formed and that a large part of the observed decrease in titer at pH 11.5 was caused by an aggregation of the virus. Electron micrographs and chemical evidence supported the observation.

Table: Inactivation of Echovirus 7 by High pH Conditions at 25 °C in a Dilution Medium.

pH	Contact time, min	Inactivation, %
11.23	7	99.98
11.49	4	99.99
11.79	4	99.5
11.92	1.5	99.9

Efficacy against Parasites

The susceptibility of protozoa or helminths to inactivation by high pH conditions does not appear to have been studied.

Mechanisms of Action

The inactivation mechanism of high pH on bacteria has not been examined. There is ample evidence to show that the poliovirus is inactivated by the disruption of the capsid and a loss of the RNA to the water. Boeye and Van Elsen suggested that, at pH 10 and at elevated temperatures ($\geq$30 °C), the RNA was released from the capsid in a degraded form or that it was quickly and extensively broken down after release. The applicability of this mechanism to enteroviruses other than poliovirus is unknown.

The utilization of high pH conditions for disinfection of water is feasible, but higher pH values than are normally used in water treatment and long contact times are required. At pH values of approximately 10.5, bacteria are inactivated in up to 600 min, but viruses are probably unaffected. A pH value of 12.0 to 12.5 and a contact time of approximately 30 min can probably yield 99.0% to 99.9% inactivation of most bacteria and of certain viruses. This process has a drawback: the pH that is necessary for effective disinfection must be reduced before the water can be consumed. Use of the high pH as the sole means of disinfection is not recommended. In many situations where water is softened by the precipitation process and where the water is of poor biological quality, the disinfection potential of this process could be used by increasing the pH above its present values.

Although this method has not been used deliberately for disinfection in the past, additional work on this method may be warranted since high pH conditions are attained in many treatment plants. Studies are needed to determine the pH values required for inactivation of protozoans, a broader and more representative group of viruses, and a larger group of enteric bacteria.

Hydrogen Peroxide

Hydrogen peroxide (H_2O_2) is a strong oxidizing agent that has been used for disinfection for more than a century. Its instability and the difficulty of preparing concentrated solutions have tended to limit its use. However, by 1950 electrochemical and other processes were developed to produce pure hydrogen peroxide in high concentration, which is known as stabilized hydrogen peroxide. This product has been subjected to increased study and application. Most recently it has been used to disinfect spacecraft, foods, and contact lenses. Although there has been some interest in using hydrogen peroxide as a disinfectant for wastewater, it has been used more for control of bulking in the activated sludge waste treatment process. Its use in drinking water disinfection appears minimal.

Analytical Methods

Analytical methods for hydrogen peroxide are based on oxidation or reduction with potassium permanganate $(KMnO_4)$, potassium iodide (KI), or ceric sulfate $\left[Ce(SO_4)_2\right]$. A colorimetric procedure based on the oxidation of titanium sulfate $(TiOSO_4)$ is sensitive to about 1 mg/liter in the absence of other oxidizing agents.

Production and Application

Commercially produced hydrogen peroxide is available in aqueous solution, usually ranging from 30% to 90%. It is stabilized during manufacture by addition of such compounds as sodium pyrophosphate $(Na_4P_2O_7)$, acetophenetidin $(CH_3CONHC_6H_4OC_2H_5)$, or acetanilide $(CH_3CONHC_6H_5)$. As the concentration of hydrogen peroxide is decreased, the concentration of stabilizers is typically increased. Experience in the waterworks industry using hydrogen peroxide is nonexistent. However, hydrogen peroxide is widely used as a bleaching agent in making cotton textiles or in wood pulping. Presumably, a concentrated solution would be diluted and applied with a chemical metering pump. Careful attention to safety in handling would be required because of the possibility of fire or explosion.

Biocidal Activity

The few pre-1965 references on disinfection by hydrogen peroxide have been summarized by Yoshpe-Purer and Eylan. Although its bactericidal activity was indicated, a need for catalytic Fe^{2+} or Cu^{2+} was reported. Yoshpe-Purer and Eylan worked with Escherichia coli, Salmonella typhi, and Staphylococcus aureus in pure culture and as a mixture of bacteria. They also studied the effect of hydrogen peroxide concentrations (from 30 to 60 mg/liter), contact time (10 to 420 min), and initial concentration of organisms. Without ever measuring residuals, they concluded that bacterial inactivations were relatively slow, that E. coli was more resistant to hydrogen peroxide than S. typhi or S. aureus, and that the required inactivation time was increased as the initial concentration of organisms was increased. Regardless of the hydrogen peroxide concentration or the type of organism used, all tests were claimed to be "sterile" in 24 hr.

Table: Contact Times Necessary for Hydrogen Peroxide (H_2O_2) to Effect 99% Inactivation of Various Concentrations of Escherichia coli, Salmonella typhi, and Staphylococcus aureus.

Test Microorganism	Initial Concentration, bacteria/ml	Contact Time Required for 99% Inactivation by Hydrogen Peroxide, min		
		30 mg/liter	60 mg/liter	90 mg/liter
Escherichia coli	10^2			~360
	10^4			
	10^6			~360
Salmonellai typh	10^2	~60	> 30 < 45	<10
	10^4	> 300	> 45 < 60	30
	10^6	< 300	> 60	> 45 < 60
Staphylococcus aureus	10^2			> 15 < 30
	10^4			~60
	10^6			~180

Toledo et al. studied inactivation of S. aureus and spores of several species of Bacillus and Clostridium with 25.8% (258,000 mg/liter) hydrogen peroxide. The time to reduce S. aureus by 6 logs was 1 min, whereas reduction of spores by 99% required from less than 1 min to approximately 17 min, depending on the species. Increasing the hydrogen peroxide concentration up to 41% (or 410,000 mg/liter) or increasing temperature from 24 °C to 76 °C significantly reduced the required inactivation time.

Several studies on virus inactivation by hydrogen peroxide have been reported. Lund obtained a 99% inactivation of poliovirus (Saukett strain) in about 6 hr with 0.3% (3,000 mg/liter) hydrogen peroxide. Mentel and Schmidt worked with rhinovirus (types 1A, 1B, and 7). They found that while a 1.5% (15,000 mg/liter) concentration required approximately 24 min for 99% inactivation, equivalent inactivation occurred in 4 min with a concentration of 3% (30,000 mg/liter).

Information is lacking on the effect of hydrogen peroxide on protozoa and helminths in water.

In none of the studies cited above is there any indication that the dosage of hydrogen peroxide was measured other than by the volumetric addition of hydrogen peroxide. No measurements of residual were reported, although Yoshpe-Purer and Eylan claimed that a residual was present for up to 13 days as indicated by inactivation of added doses of bacteria.

Mechanism of Action

No studies have specifically identified the mechanism of action of hydrogen peroxide. Spaulding et al. believed that the hydrogen peroxide molecule itself was not responsible for the action but, rather, that the free hydroxyl radical ($HO \cdot$) that it produced was the specific inactivating agent. They claimed that the catalytic effect of iron or copper ions supported this theory. Yoshpe-Purer and Eylan also reported that free radicals were important but that sufficient catalyzing metal ions were available either from the tap water that they used or from the bacterial cells themselves.

Because of its relatively high cost and the high concentrations that are required to achieve disinfection in reasonable time, hydrogen peroxide is not a generally satisfactory disinfectant for drinking water.

If further research is to be conducted, the stability of hydrogen peroxide and its residual effect in the presence of organic material and other substances in water or distribution systems should be investigated. Parallel studies using chlorine and hydrogen peroxide, separately and in conjunction, should be conducted.

Ionizing Radiation

Ionizing radiation may be electromagnetic or particulate. As used for disinfection or sterilization, electromagnetic radiation may be UV, gamma, or X rays, and the particles may be alpha or beta or neutrons, mesons, positrons, or neutrinos. This discussion is limited to gamma rays and beta particles (or electrons).

Although there is extensive literature on the use of ionizing radiation in the preservation of food and other materials, little information is available on water treatment.

Application

Laboratory studies of destruction of microorganisms are generally conducted by exposing small containers holding the test suspensions to the shielded source of ionizing radiation. However, in a practical application, the shielded source must be designed to permit relatively thin sheets of flowing liquid to be exposed. This is particularly important for high-energy electrons that have a less penetrating power than gamma rays. The MIT study included a field demonstration in which a plant treating approximately 380,000 liters/day (minimum dosage of 400,000 rads) was designed and operated. The energy source was a 50-kW electron accelerator operated at 850,000 V. Sludge was pumped under the electron beam through a drum system that produced a layer that was 1.2-m wide and 2-mm thick. The entire operation was conducted in a concrete vault to shield the operators against X rays, which are produced incidentally. Shielding of workers is a major requirement both for gamma ray and high-energy electron sources.

Analytical Methods and Residual

No residual is produced by ionizing radiation. Consequently, dosage has been measured exclusively. The MIT report and Silverman and Sinskey summarized analytical methods including calorimetry, photoluminescent dosimetry, colorimetry, and the use of ionization chamber instruments.

Biocidal Activity

Efficacy against Bacteria

The first study of ionizing radiation in the treatment of water was reported by Dunn. In addition to reviewing the literature and providing some discussion on ionizing radiation, he studied the use of a 1,000-Ci source of cobalt-60 (providing gamma rays) and a Van de Graaff generator (providing high-energy electrons). Working with natural waters and wastewaters that were not chemically or bacteriologically characterized or controlled, he exposed samples to a constant radiation flux (2,000 R/min from the cobalt-60 or electrons from 3-MV operation of the Van de Graaff generator). Varying only the time of exposure, Dunn found that it took 0.125 min (dosage 250,000 R) with the cobalt-60 source to reach 95.4% to 99.999% inactivation of total initial numbers of bacteria (as measured by a plate count at 37 °C).

Ridenour and Armbruster studied the effect of a 10-kCi source of cobalt-60 (3,000 R/min) on a variety of natural waters, wastewaters, and pure cultures (approximately 2×10^7 organisms per milliliter) of 10 different organisms. They found that a dosage of 100,000 roentgen equivalent physical (rep) reduced the count of all species tested by > 99%. (In order of increasing resistance, the species were: Enterobacter aerogenes, Escherichia coli, Shigella flexneri, Salmonella typhi, S. sonnei, Salmonella sp., Staphylococcus aureus, S. paratyphi B, Streptococcus faecalis, and Bacillus subtilis.) With river water, a dosage of from 50,000 to 100,000 rep reduced the total bacterial count and the coliform index by at least 99%, but 150,000 rep were required to reduce the streptococcus index by 99%. Varying the pH (5.0, 7.0, and 8.5) had no effect on the rate of inactivation.

Also using cobalt-60 (1,100 Ci), Lowe et al. exposed pure cultures of bacteria that were suspended in double distilled water and sterile settled sewage. The concentration of microorganisms was approximately 1×10^6 / ml. Although they did not report exposure time, they summarized their exposure data in rads (radiation absorbed dose).

Table: Dosages of Cobalt-60 Irradiation Necessary for 99% Inactivation of Microorganisms in Distilled Water.

Test Microorganism	Dosage Required, rad
Bacillus subtilis var niger	3.5×10^5
Mycobacterium smegmatis	1.4×10^5
Escherichia coli	6.5×10^4
Microcccus pyogenes var aureus	5.8×10^4
E. coli phage T3	3.2×10^4

There have been more studies on the disinfection of wastewater than on disinfection of drinking water. Ballantine et al. and Compton et al. considered ionizing radiation not as good as other available methods but, with increased emphasis on water reclamation and the use of high-energy electrons, the prospects for practical application seem improved. In the MIT study, which is the most systematic one conducted to date, washed cells or spores were suspended in 0.067-M phosphate buffer (1×10^8 cells/ml) and exposed to high-energy electrons from a Van de Graaff generator. The approximate dosages, in rads, required for a 99% inactivation of the organisms tested were: Escherichia coli (K12), 38,000; Salmonella typhimurium (LT2), 24,000; Micrococcus sp., 35,000; Aspergillus niger (spores), 78,000; S. typhimurium (24), 100,000; Streptococcus fecalis, 300,000; and Clostridium perfringens (spores), 400,000. The investigators found that the solids in the wastewater had no effect on disinfection.

Of the factors affecting disinfection, oxygen was most important. For example, for E. coli K12, the dosage required for 99% inactivation was more than doubled when an atmosphere of air was replaced by one of nitrogen, and almost halved when the air was replaced by an oxygen-rich atmosphere.

Efficacy against Viruses

The inactivation of viruses does not appear to be influenced by the source of the ionizing radiation.

It is affected by factors that are similar to those described above for bacterial inactivatio. Lowe et al. found that 32,000 rads were required for 99% inactivation of E. coli phage T3. In the MIT study, from 300,000 to 420,000 rads were required for 99% inactivation of coxsackievirus B3, poliovirus 2, echovirus 7, reovirus 1, and adenovirus 5 (in order of increasing sensitivity), each suspended in 0.05 M glycine at pH 7.0.

Efficacy against Parasites

The MIT study appears to be the only one in which the effect of ionizing radiation on both protozoa and helminths was investigated; however, no quantitative data were reported. Brannan et al. found that to attain a 90% reduction of embryonation of Ascaris lumbricoides eggs, a dosage of approximately 30,000 rads was required.

Mechanism of Action

In a review of the literature, Silverman and Sinskey summarized the mode of action of disinfection by ionizing radiation. Two effects are recognized: the direct effect, in which the primary cellular target is DNA that is damaged by the energy released from the ionizing radiation, and the indirect effect, which is associated with the production of such substances in the cell menstruum as hydrogen peroxide (H_2O_2), organic peroxides, and free radicals. Oxygen is important because it reacts with electrons and radicals, which in turn react to form hydrogen peroxide or ozone (O_3).

Ionizing radiation can disinfect water effectively; however, the large source, shielding, and relatively thin exposure layers that are required create difficult engineering and safety problems. The complex technology may limit application to large facilities that can provide for adequate safeguards. The absence of residual disinfection is also restrictive.

Potassium Permanganate

Potassium permanganate $(KMnO_4)$ is a strong oxidizing agent, which was first used as a municipal water treatment chemical by Sir Alexander Houston of the London Metropolitan Water Board in 1913. In the United States, it was used in Rochester, New York, in 1927 and in Buffalo, New York, in 1928. Since 1948, it has been used more widely in waterworks as an algicide, as an oxidant to control tastes and odors, to remove iron and manganese from solution, and, to a limited extent, as a disinfectant.

The relatively limited information concerning disinfection with potassium permanganate is subject to criticism, because there have been no studies of the effects of organic constituents of the medium (test system) or destruction of a variety of organisms, especially pathogens.

Analytical Methods and Residual

Concentrations of potassium permanganate can be determined readily by iodometric titration using sodium thiosulfate $(Na_2S_2O_3)$ as a titrant or by direct titration with ferrous sulfate. Manganese can be determined by atomic absorption spectrophotometry, or the permanganate ion (MnO_4^-) can be measured colorimetrically. The minimum detectable concentration of manganese by the colorimetric procedures, using a 100-ml sample, is 50 μg/liter. A concentration of 0.05 mg/liter or greater can be detected visually by the pink color that is imparted to the water.

Production and Application

Crystalline potassium permanganate is highly soluble in water (2.83 g/100 g at 0 °C). In water-works, it is prepared usually as a dilute solution (1% to 4%) and applied with a chemical metering pump. It also may be added as a solid using conventional dry-feed equipment.

Reduction of the permanganate ion produces insoluble manganese oxide (Mn_3O_4) hydrates. To prevent distribution of turbid water or of water that will cause unsightly staining of plumbing fixtures, potassium permanganate most often is applied as a pretreatment that is followed by filtration. For example, addition of potassium permanganate to a finished water to maintain a residual in a distribution system is unacceptable because of the pink color of the compound itself or the brown color of the oxides.

Biocidal Activity

Cleasby *et al.* and Kemp *et al.* summarized the few scientific references to disinfection by potassium permanganate that were published earlier than 1960. Although some bactericidal activity was indicated, no quantitative data were presented, and the early reports disagreed as to its value.

The most systematic study of disinfection by potassium permanganate has been made by Cleasby *et al.*. They worked exclusively with *Escherichia coli* (prepared as a lactose broth culture) and studied the effect of potassium permanganate at doses of 1 to 16 mg/liter at pH values of 5.9, 7.4, and 9.2, temperatures of 0 °C and 20 °C, and contact times of 4 to 120 min. They concluded that bacterial inactivation was relatively ineffective, but slightly better at the higher temperature, and that increasing pH decreased disinfection rates. Table summarizes in tabular form their data, which were presented originally in graphic fashion.

Table: Concentrations and Contact Times Necessary for Potassium Permanganate to Effect 99% Inactivation of a 48-Hr Escherichia coli Lactose Broth Culture.

Contact Time Required for 99% Inactivation by Potassium Permanganate, min							
1 mg/liter	2 mg/liter	4 mg/liter	8 mg/liter	12 mg/liter	16 mg/liter	pH	Temperature, °C
45				10	5	5.9	0
					115	7.4	
						9.2	
95	45	15	15	10	5	5.9	20
			80	80	25	7.4	
						9.2	

At least two patent applications involve the use of potassium permanganate as a disinfectant for water in swimming pools. Seidel's patent covered use of a concentration of 0.1 to 0.2 mg/liter in a recirculating system including filtration through quartz gravel. She claimed removal of bacteria and algae, but the specific merit of potassium permanganate in her system cannot be determined from the patent application. Heuston proposed a tablet containing 0.001 g of potassium permanganate per 0.2 g tablet to give a dose of 1 mg/liter. Because the tablet includes potassium iodide (KI) and other oxidizing agents, it is impossible to assess the specific killing action of potassium permanganate.

A number of studies on inactivation by potassium permanganate have been reported. However, these dealt largely with the mode of action of potassium permanganate in viral inactivation or at efforts to control human diseases, animal diseases, or plant diseases. They provide few clear-cut quantitative data on biocidal activity of the compound.

Information is lacking on the effect of potassium permanganate on protozoans and helminths in water.

Mechanism of Action

The mechanism of action of potassium permanganate has not been definitively identified. From the studies that have been conducted it may be presumed that it exerts its disinfection activities by oxidizing compounds that are involved in essential cellular functions. However, Lund questioned the suggestion that the oxidative process was the mechanism for poliovirus inactivation, but was unable to exclude it as a possibility.

The relatively high cost, ineffective bactericidal action, and aesthetic unsuitability of maintaining a residual in the distribution system make potassium permanganate a generally unsatisfactory disinfectant for drinking water.

Additional studies on potassium permanganate are hardly warranted, but in the unlikely event that further research on its biocidal activities is conducted, it would be desirable to study the effect of organic material on its disinfection and the effect of the compound on organisms other than E. coli. To relate disinfection efficiency to more conventional practices, parallel studies using chlorine should be conducted.

Silver

Silver as a metal has been known for millennia, and its use as a water disinfectant dates back to the Persian king Cyrus. The term oligodynamic, which describes the killing effect of small concentrations, was coined by von Naegeli in 1893, but it is unscientific and its use should be rejected. The antibacterial action of silver and silver nitrate $(AgNO_3)$ was noted first by Raulin in 1869. Since the late nineteenth century, but especially since World War II, there have been considerable efforts to exploit the use of silver as a disinfectant, particularly for individual (home) water systems and swimming pools. Silver has been used both as a salt, most commonly silver nitrate, or as metallic silver, either bound in filter beds, generated by electrolytic devices, or applied as a colloidal suspension.

The relatively limited information concerning disinfection with silver has been seriously criticized because accurate measurement of low concentrations, i.e., < 200 μg/liter, has been difficult; a suitable neutralizing agent has not always been incorporated in test protocols; and adsorption of silver on surfaces of test vessels has confounded some studies. Despite these limitations, silver has been used in parts of Europe and Japan as a water disinfectant.

Production and Application

In water treatment, silver has been applied principally by dissolving the metal or by incorporating a silver compound in a filter medium, often an activated carbon filter. Romans has described older processes and patents. Davies reviewed processes for treating swimming pool water in which silver was added as a soluble salt and the silver ions were kept oxidized by the addition of persulfate; metallic silver was deposited on an activated carbon filter; silver was released from solid silver electrodes, which were alternately made anodic and cathodic; silver was released from an activated carbon filter containing silver and supplemented by passing the water through anodic silver screens; and silver was released from a pair of silver-copper alloy electrodes by applying a low reversing voltage.

During the past 6 yr, dozens of patents have been issued in Germany, Spain, South Africa, India, and the United States for devices or systems for adding silver to home drinking water systems or swimming pools. The home systems are most frequently combinations of activated carbon to remove tastes and odors and silver to prevent bacterial growth on the filter.

Because of the low solubility of silver, the dose is usually less than 50 µg/liter. Presumably this is reduced rapidly because of the adsorption of silver to surfaces, but there is no information on the silver residual in treated water. However, the maximum contaminant level (MCL) is 0.05 mg/liter.

Analytical Methods

Until recently there have been no satisfactory techniques for measuring silver at the µg/liter level. Using a dithizone colorimetric method, the minimum detectable quantity of silver is 200 µg/liter. Using an atomic absorption spectrophotometric method, the detection limit is 10 µg/liter, and, if a heated graphite furnace is used, the detection limit is reduced to 0.005 µg/liter.

Biocidal Activity

One of the earliest studies was conducted by Just and Szniolis, who found that 100 µg of silver per liter disinfected water within 3 to 4 hr, that silver nitrate and metallic silver were equally effective, and that histopathological changes occurred in rats that were given water containing 400-1,000 µg silver/liter for 100 days. Other systematic studies of silver disinfection have been conducted by Renn and Chesney, Wuhrmann and Zobrist, and Chambers and Proctor. These studies have been summarized by Woodward. Table shows that in the concentrations used, silver acted slowly, but that there were increases in the rate of bacterial inactivation with increasing temperature and pH. Chambers and Proctor obtained similar results, but the inactivation rates of Renn and Chesney, who used comparable concentrations of silver, were faster.

Table: Concentrations and Contact Times Necessary for Silver to Effect 99.9% Inactivation of Escherichia coli in 0.005 M Phosphate Buffer.

Contact Time Required for 99.9% Inactivation by Silver as Silver Nitrate, min					
0.01 mg/liter	0.03 mg/liter	0.09 mg/liter	0.27 mg/liter	pH	Temperature, °C
1,010	837	156	53	6.3	5
466	214	81	34	7.5	
268	109	58	18	8.7	
831	344	144	32	6.3	15
316	177	63	21	7.5	

216	100	38	13	8.7	
1,210	152	68	20	6.3	25
423	86	32	13	7.5	
203	40	20	8	8.7	

Both Wuhrmann and Zobrist and Chambers and Proctor found that the presence of phosphate (at 60 mg/liter) slowed bacterial inactivation. Because phosphate is typically found at much lower concentrations in drinking water, this observation has little practical significance, but it may account for disparate results in some laboratory studies.

Increasing water hardness slowed bacterial inactivation. According to Wuhrmann and Zobrist, an increase of 3 min was required to achieve 99.9% bacterial inactivation (at 20 °C and pH 7.0) for each 10 mg/liter increase in hardness. Likewise, chlorides interfered with the action of silver. At 10 mg/liter, they increased the inactivation time by 25% and at 100 mg/liter, by 70%.

The source of silver seemed to be irrelevant to the inactivation rate because silver either added as silver nitrate or dissolved from metallic silver gave approximately equal results.

In most studies of the disinfection of water with silver, *Escherichia coli* has been the test organism. Wuhrmann and Zobrist also tested a *Salmonella* species and found it to be at least as sensitive as *E. coli*. Yasinskii and Kuznetsova observed that 90 µg/liter inactivated *Vibrio comma* $(1\times10^6 \text{ cells}/\text{ml})$ in 30 min or 22.5 µg/liter in 60 min.

Fair and Harrison suggested that silver salts were ineffective against cysts of *Entamoeba histolytica,* but the data provided in these reports were limited. By contrast, Newton and Jones observed that electrolytically produced silver in tap water gave greater than 99% cyst inactivation in 1 hr and highly variable residual concentrations that ranged between 17 and 33 mg/liter at pH's of 9.0 to 9.8. When silver nitrate in distilled water at 0.5 mg/liter was used, 4 to 6 hr were required for a 90% to 99% inactivation. At 5 mg/liter, a > 99% inactivation occurred in 3 hr, and at 30 mg/liter, a > 99% inactivation occurred in I to 2 hr.

Chang and Baxter used 150 mg/liter of silver nitrate (95 mg/liter Ag$^+$) with contact lines of 1 to 6 min. They concluded that cysticidal activity was only moderate compared to that of iodine. Apparently, there are no data on silver as a virucidal agent in water except for one study by Lund (1963), who decreased poliovirus activity by 2.5 logs in 4 hr with a concentration of approximately 68 mg/liter.

A good general review of silver and its compounds and a more thorough medical examination of their applications was published by Grier, who concluded that there would be a "significant place for silver compounds in the prevention and treatment of at least some bacterial diseases."

Mechanism of Action

Romans summarized the mechanism of so-called oligodynamic activity: "However, there is much difference of opinion as to the form of the active principle, as to the mechanism of its action, and the value of the results obtained. Some authorities believe that the active form is a positively charged ion, some think it is a complex ion, others a salt and still others think it acts by formation of proteinates or merely as a catalyst. These are only a few of the ideas that have been expressed with formidable experimental support."

Chang attributed a direct action to silver in the nonreversible formation of silver-sulfhydryl complexes that could not function as hydrogen carriers:

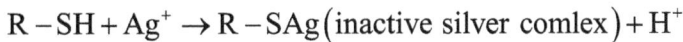

$$2R-SH \rightarrow R-S-S-R \ (normal\ sulfhydryl) + H_2$$

$$R-SH + Ag^+ \rightarrow R-SAg(inactive\ silver\ comlex) + H^+$$

He considered silver to be bacteriostatic as well as bactericidal. This would explain the relatively long contact times that are required for antibacterial activity at the concentrations that are normally used.

Zimmerman demonstrated that low concentrations of silver do not enter the cell but are adsorbed onto the bacterial surface just as silver tends to be adsorbed on other surfaces. According to Chang, the adsorbed silver ions must immobilize the dehydrogenation process because bacterial respiration takes place at the cell surface membrane.

As a disinfectant, silver may be applicable to home treatment systems and swimming pools. These applications may be more effective in keeping filters free from bacteria than in actually inactivating organisms that are suspended in the water. This is because of the relatively low rate of solution of silver in water and its low biocidal activities.

Woodward ably summarized the situation: "The high cost of silver will limit it to specialty uses. The fact that silver does not impart taste, odor, or color to water makes it attractive for use. Its slow bactericidal action, although a disadvantage in some situations, may be an advantage in others, particularly where water is stored for long times before use, as on shipboard. Until some of the uncertainties regarding silver are resolved, it would be prudent to use it as a drinking water disinfectant only in situations where substantial factors of safety can be provided and where the bactericidal effectiveness of the procedure can be monitored."

Silver and its compounds are weak, costly disinfectants that are unsuitable for use in municipal drinking water supplies. To achieve acceptable disinfection in a reasonable time would require concentrations exceeding the MCL of 0.05 mg/liter.

Now that adequate chemical analytical techniques are available for measuring low concentrations of silver, it would be desirable to conduct disinfection studies in which both dose and residual are measured accurately. To relate disinfection efficiency to more conventional practices, parallel studies using chlorine should be conducted.

Ultraviolet Light

Electromagnetic radiation, in wavelengths from 240 to 280 nm, is an effective agent for killing bacteria and other microorganisms in water. Conveniently, from a practical point of view, from 30% to more than 90% of the energy emitted by a low-pressure mercury arc, which is enclosed in special UV transmitting glass, is emitted at a wavelength of 253.7 nm.

Two basically different physical arrangements are commonly used for the application of UV light to water. In one, lamps are placed above the solution to be disinfected at the apex or focus of parabolic or elliptical reflectors. For this purpose, aluminum is preferred because of its high reflectance for the germicidal 253.7-nm wavelength. Although this is an efficient way of applying UV radiation to water, the open nature of the structure can permit contamination. Furthermore, it must operate at atmospheric pressure. Tubular reactors are more common in water treatment because they are sealed and operate under pressure.

To increase intensities and permit higher flow rates, multiple lamp reactors are being designed and put into use. These units generally contain lamps that are positioned parallel to the flow of water through them. A recent patent describes an apparatus using a water film approximately 0.64 cm thick and a flow of liquid that is perpendicular to the lamp in a baffled system. This type of unit is designed for disinfection of fluids with low UV transmittance. For liquids with high transmittance like drinking water, such a design is inefficient, because insufficient depth is available for absorption of UV and a large fraction of the radiant energy is dissipated as heat when the UV is reflected on the walls of the contactor.

In units containing lamps that are surrounded by water, an insulation space must be provided to maintain their efficiency. The maximum efficiency of modern cold cathode lamps is near 40 °C, dropping off to a 50% output at 24 °C and 65 °C. Lamps are normally placed in sleeves that are made of high-silica glass or quartz in order to maximize transmission at the 253.7-nm wavelength. Solarization or opacity can develop in the sleeve as it deteriorates with age. The sleeves must be cleaned regularly for efficient functioning.

Another consideration in the mechanical design of UV contactors is the degree of agitation that is provided and the plug flow characteristics of the contactor. In all disinfection systems that require several orders of magnitude of disinfection, short circuiting or nonplug flow characteristics of contactors, which bypass a part of the flow, limit the efficiency of the process. It was recognized by Cortelyou et al. in 1954 that the degree of agitation in the UV disinfection of water is important in bringing the target microorganism into close proximity to the UV source where the intensity is highest.

Dose

The dose, D, of electromagnetic radiation that is applied to a solution is commonly measured as the intensity of radiant energy input, I_0, at the lamp surface or at some given distance from the lamp (I_0 being expressed as $\mu W / cm^2$), multiplied by the time of exposure, t, in seconds:

$$D = I_0 \, t \, \mu W \cdot s / cm^2$$

Another measurement of dose is chemical actinometry. In this method, a photochemical reaction with a known quantum yield is used to measure the intensity (quanta per second) of light that is absorbed by the actinometry solution. Knowing the time of exposure, the volume of solution, and the volume of sample analyzed, the moles of photochemical reaction per unit of time can be related through the quantum yield to the Einsteins or moles of photons per second that are absorbed by the fluid. From the wavelength of radiation, l, and the area of lamp surface, the average intensity at the lamp surface can be determined.

$$E = hc / \lambda = 1.2 \times 10^7 / \lambda \; W \cdot s / Einstein \; nm$$

Neither of these measurements of energy input, I_0, per unit area of lamp exposed considers the change in intensity through the depth of exposed fluid and the solid angle over which this energy interacts. Recently, attempts have been made to correct the decrease in intensity with depth in studies of wastewater disinfection where this problem is acute. These studies have shown that the usual water quality indices, such as chemical oxygen demand (COD), color, and turbidity, do not adequately predict the loss of intensity through the solution. The direct measurement of UV transmittance has been successful. In spite of these difficulties, Huff et al. suggested that-color at a maximum level of 5 units, iron at 3.7 mg/liter, and turbidity up to 5 units did not decrease treatment efficiency below acceptable limits in a unit with a maximum water depth of approximately 7.5 cm.

Biocidal Dose

The biocidal dose of UV energy consists of the intensity of UV energy that is absorbed at the reactive site within the microbe over the time of interaction. The biocidal dose is then a function of the energy input from the UV source into the solution, dispersion of the energy as a function of distance from the source, the depth of the fluid between the organisms and the source as well as its absorptivity, and, finally, the losses and reflection of UV light within the contactor. All of these factors determine the actual intensity of radiation that is available to the microorganism at any one point within the contactor.

Quite early in the study of the batericidal action of UV light, Gates found that the relation between the intensity of incident energy and time required for bacterial destruction were not of equal importance in determining experimental disinfection. However, Hoather found that when one corrects for the transmitted intensity of radiation, taking into consideration the absorptivity of the water and dispersion as a function of distance from the source, the required time of exposure is inversely proportional to the calculated intensity of radiation penetrating the water for a given degree of inactivation. Finally, Oliver and Cosgrove used chemical actinometry and a pulsed laser to show that their coliform and streptococcal inactivation in wastewater depended only on the total dose that is delivered and not on the dose rate or intensity of light that impinged on the sample. Thus, the same number of μW s / cm^2 provided the same disinfection regardless of the time and intensity that were used to produce that dosage.

Bacterial spores have also been studied with chemical actinometry. For certain stocks, the rate at which the energy was deposited affected the degree of response. Powers and his colleagues observed that the effect of the dose rate was not due to geometric factors or intensity variations with depth and solution absorptivity. This explains many earlier results. They did find that dependence on dose rate was seen only for some of their stocks of spores, it was not as dramatic with monochromatic-filtered light of a wavelength of 253.7 nm, and it was observed only at very low doses. The explanations usually given for these low-dose effects are photochemical back reactions, enzymatic repair, and photoreactivation mechanisms.

Residual

UV radiation produces no residual. Therefore, monitoring and control of disinfection efficiency are more difficult for UV than for chemical disinfectants.

However, this is not a major problem, because disinfection can be controlled by adjusting the contact time and the UV energy that is transmitted through the solution. The major disadvantage is the lack of a tracer for ensuring the integrity of the distribution system.

Biocidal Activity

UV disinfection follows Chick's Law of kinetics:

$$-\log N / N_0 = \frac{I \times t}{Q}$$

Where, N is the number surviving at time t of an original population N_0 that has been exposed to an intensity I. Q is the dose for 1 log survival, one lethal unit exposure, or a lethe.

The variables discussed above interfere with the accurate determination of the average intensity and time of exposure in a practical UV contactor. Consequently, there is considerable disagreement in the literature concerning the absolute magnitude of Q for the microorganisms that have been studied.

Of the Gram-negative bacteria, Escherichia coli is consistently more resistant to disinfection by UV than are the Salmonella and Shigella species that have been studied. The vegetative forms of the Gram-positive bacteria were more resistant than E. coli. Kawabata and Harada found Streptococcus faecalis nearly 3 times more resistant than E. coli, while Bacillus subtilis was 4 times more resistant. The spores of B. subtilis were 6 times more resistant than E. coli.

Luckiesh and Holladay reported that a dosage of 2,400 $\mu W \cdot s / cm^2$ at a wavelength of 254 nm produced one lethe of disinfection of E. coli. Huff et al. studied a two-lamp shipboard disinfection unit with an approximately 75-cm long, 20-cm diam. unbaffled stainless steel cylinder. Their dosages were generally quite high, ranging from 3,000 to 11,000 $\mu W \cdot s / cm^2$ at a maximum water depth of approximately 19 cm. At the lowest dosage, there was a 0.02% to 0.04% survival of E. coli, which, assuming Chick's Law, gives a dosage of less than 400 $\mu W \cdot s / cm^2$ / lethe. They also found approximately 2 logs less spore inactivation with the same dose as required for vegetative cells. They reported more than 4 logs of inactivation of polio-, echo-, and coxsackievirus with 4,000 $\mu W \cdot s/cm^2$. Based on these results, the U.S. Department of Health, Education, and Welfare issued a policy statement on April 1, 1966, stating criteria for the acceptability of UV disinfecting units as a minimum dosage of 16,000 $\mu W \cdot s/cm^2$ with a maximum water depth of approximately 7.5 cm.

Morris has compared the dose of 254-nm UV energy required to inactivate a range of microorganisms suspended in droplets of buffer solution on an aluminum surface. Their results are shown in table expressed as $\mu W \cdot s / cm^2$ / lethe.

Table: Ultraviolet Energy Necessary to Inactivate Various Organisms.

Test Microorganism	Lethal Dose, ($\mu W \cdot s/cm^2$)
Escherichia coli	360
Staphylococcus aureus	210
Serratia marcescens	290
Sarcina lutea	1,250
Bacillus globiggii spores	1,300

T3 coliphage	160
Poliovirus	780
Vaccinia virus	30
Semliki Forrest virus	470
EMC virus	650

Required dosages reported by Morris are similar in magnitude to those found by Huff et al. Both Huffs values and those of Morris are lower than those of Luckiesh and Holladay, partly because of the additional dose of UV that was provided by reflections from the contactor and aluminum surface, which produce an estimated dosage approximately twice the values given. No studies of the action of UV light on parasitic helminths and protozoa in the treatment of drinking water appear to have been reported.

Mechanism of Action

Inactivation by UV light is believed to act through the direct absorption of UV energy by the microorganism, causing a molecular rearrangement of one or more of the biochemical components that are essential to the organism's functioning. The major site of UV absorption in microorganisms is the purine and pyrimidine components of the nucleoproteins. Since the relative efficiency of disinfection by UV energy as a function of wavelength follows the absorption spectrum of these chromophore groups and because the first law of photochemistry states that the light that is absorbed by a molecule can produce photochemical change, the mechanism of action of UV probably occurs through the progressive biochemical change that is produced primarily in the nucleoproteins. Witkin has reviewed both UV mutagenesis and inducible DNA repair. The major specific mechanisms that have been suggested for UV damage involve the reversible formation of pyrimidine hydrates and pyrimidine dimers. Breaks in the bonding structure, known as nicking, also occur.

The repair of damaged nucleoprotein with light of a wavelength longer than that of the damaging radiation is commonly referred to as photoreactivation. This process, originally identified by Kelner, has been studied extensively. It occurs in the visible wavelength range from 300 to 550 nm. In addition, repair of nucleoproteins that have been damaged by UV radiation can also occur in the dark. The mechanisms of repair of damage to DNA caused by UV radiation are generally thought to involve induced enzymes, but repair has also been observed in inactive systems for which photochemical back reactions have been suggested.

Drinking Water Treatment

Precipitation and Coagulation

Precipitation is a technique of removing one or more substances from a solution by adding reagents so that insoluble solids appear. The 'solubility' rules the technique, i.e., when the product of ion concentrations (in simple) in the solution over the solubility product of the respective solid, the precipitation occurs. It is one of the simple methods to purify water. The chemicals are added to form particles which settle and remove contaminants from water. The treated water is reused whereas the settled portion is dewatered and disposed of. The technique is used in softening of water as well as to remove impurities like phosphorus, fluoride, arsenic, ferrocyanide and heavy metals, etc.

Softening of Water

The presence of Ca/Mg in terms of carbonate, bicarbonate, chloride and sulfate results in hardness of water. Addition of proper chemical forms precipitation and makes it soft.

Addition of $Ca(OH)_2$ forms precipitation with bicarbonate and sulfate in water.

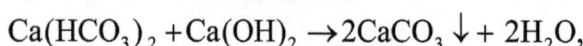

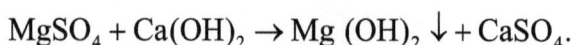

$$Ca(HCO_3)_2 + Ca(OH)_2 \rightarrow 2CaCO_3 \downarrow + 2H_2O,$$

$$MgSO_4 + Ca(OH)_2 \rightarrow Mg(OH)_2 \downarrow + CaSO_4.$$

Addition of Na-aluminate forms precipitation of hydroxide with sulfate and chloride in water. Actually, Na-aluminate forms sodium hydroxide with water, and with sulfate/chloride it forms hydroxide.

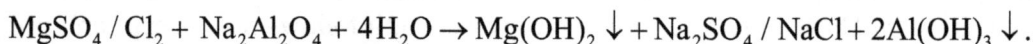

$$MgSO_4 / Cl_2 + Na_2Al_2O_4 + 4H_2O \rightarrow Mg(OH)_2 \downarrow + Na_2SO_4 / NaCl + 2Al(OH)_3 \downarrow.$$

Formation of aluminum hydroxide aids in floc formation, sludge blanket conditioning and silica reduction.

Softening of water is also feasible by simple boiling:

$$Ca(HCO_3)_2 + heat \rightarrow CaCO_3 \downarrow + H_2O + CO_2.$$

Removal of Heavy Metals

Heavy metals (e.g., Ba, Cd, Pb, Hg, Ni, Cu) typically precipitated from waste water as sulfates, sulfides, hydroxides, and carbonates. Metal co-precipitation during flocculation with iron and aluminum salts is also possible for some metals (e.g., As, Cd, Hg, Cr). The following reaction represents as chromium co-precipitation in terms of hydroxides or sulfates.

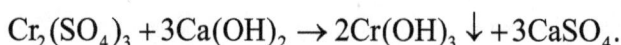

$$H_2Cr_2O_7 + 6FeSO_4 + 6H_2SO_4 \rightarrow Cr_2(SO_4)_3 \downarrow + 3Fe_2(SO_4)_3 + 7H_2O,$$

$$Cr_2(SO_4)_3 + 3Ca(OH)_2 \rightarrow 2Cr(OH)_3 \downarrow + 3CaSO_4.$$

Removal of Arsenic

Arsenic removal with coagulants, viz. Alum $\left[Al_2(SO_4)_3 \cdot 18H_2O\right]$ ferric chloride $(FeCl_3)$ and ferric sulfate $\left[Fe_2(SO_4)_3 \cdot 7H_2O\right]$ is effective. In these cases, arsenic (V) can be more effectively removed than arsenic (III). The microparticles and negatively charged arsenic ions are attached to the flocs by electrostatic attachment during the process. The possible steps of coagulation and co-precipitation are as follows:

Alum dissolution:

$$Al_2(SO_4)_3 \cdot 18H_2O \rightarrow 2Al^{+3} + 3SO_4^{-2} + 18H_2O.$$

Aluminum precipitation (acidic):

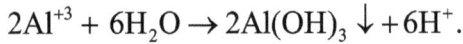

$$2Al^{+3} + 6H_2O \rightarrow 2Al(OH)_3 \downarrow + 6H^+.$$

Co-precipitation (non-stoichiometric, non-defined product):

$$H_2AsO_4^- + Al(OH)_3 \rightarrow Al - As \, (complex) + Other \, Products.$$

Similar reactions take place in case of ferric chloride and sulfate with the formation of Fe–As complex as an end product which is removed by the process of sedimentation and filtration. The efficient removal depends on pH range.

Removal of Phosphorus

The removal of phosphates is generally done by coagulant, i.e., by mixing coagulant into waste water. The most commonly used multivalent metal ions are Ca, Al, and Fe.

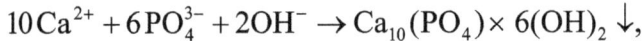

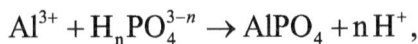

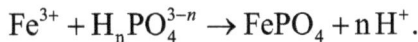

$$10Ca^{2+} + 6PO_4^{3-} + 2OH^- \rightarrow Ca_{10}(PO_4) \times 6(OH)_2 \downarrow,$$
$$Al^{3+} + H_nPO_4^{3-n} \rightarrow AlPO_4 + nH^+,$$
$$Fe^{3+} + H_nPO_4^{3-n} \rightarrow FePO_4 + nH^+.$$

Removal of Fluoride

Precipitation of fluoride species into chemically stable form is the most effective option for the removal of fluoride (in terms of Ca, Mg, Al) from effluent streams. Among all metal fluorides, CaF_2 is less soluble in water. Consequently, removal of fluoride from the effluents by converting it into CaF_2 has become the most widely used method of treatment. $CaCl_2$, limes, may be used for this purpose, but $CaCl_2$ is preferred with respect to lime due to its higher solubility and the lower ratio of additive to effluent.

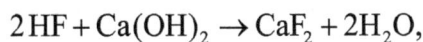

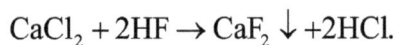

$$2HF + Ca(OH)_2 \rightarrow CaF_2 + 2H_2O,$$
$$CaCl_2 + 2HF \rightarrow CaF_2 \downarrow + 2HCl.$$

The reaction of hydrofluoric acid and ammonium fluoride with the aluminum treatment agent is as follows:

$$3HF + AlO_2 \rightarrow AlF_3 \downarrow + H_2O + OH^-,$$
$$3NH_4F + AlO_2 \rightarrow AlF_3 \downarrow + 3NH_4^+ + O_2,$$
$$6NH_4HF_2 + 4AlO_2 \rightarrow 4AlF_3 \downarrow + 6NH_4^+ + 2O_2 + 2H_2O + 2OH^-.$$

Inorganic flocculants have the potential in different separations, but they are used in very large quantities. These leaves large amounts of sludge and strongly affected by pH changes, whereas polymeric flocculants cause the formation of large cohesive aggregates (flocs) and inert to pH changes. Both natural and synthetic polymers are useful for this purpose. Generally, synthetic polymers (viz. polyacrylamide, polyethylene oxide, poly (diallyl dimethyl ammonium chloride), poly (styrene sulfonic acid) are highly effective flocculants at small dosages and have high tailor ability but poor shear stability, whereas though natural polymers (viz. starch, guar gum, alginate, glycogen, dextran) are biodegradable and effectively shear stable.

Removal of Dyes

Dyes are non-biodegradable, and precipitation with $CaCO_3$ can be one of the approaches to remove them from the water.

Benefits:

- Simple process,

- Effective for the removal of As, Cd, Ba, Cd, Cr, Pb, Hg, Se, Ag, etc.,

- It is also applicable to remove natural organic matter (NOM) or dissolved organic carbon (DOC).

Limitations:

- Requires continuous supply of huge chemicals,

- Handling of by-products,

- Disposal of coagulation/precipitation sludge is a concern.

Distillation

It is the most common separation technique. In this separation technique, the mixed components in water are separated by the application of heat. It is based on the differences in boiling points of the individual components. The boiling point characteristics depend on the concentrations of the components present. Thus, the distillation process depends on the vapor pressure characteristics of liquid mixtures. The basic principle described as the input of heat energy raises vapor pressure. When the vapor pressure reaches its surrounding pressure, the liquid mixture boils and distillation occurs because of the differences of volatility in the mixture.

This process results in a separation between water and inorganic substances, such as lead, calcium, magnesium, etc. are also destroying bacteria. However, organics with boiling points lower than 100 °C cannot be removed efficiently and can actually become concentrated in the product water. Distilled water purification technology was originally developed for industrial purpose. However, it came eventually for home use. Since, this process is not very effective in removing organic chemicals so the carbon filter system must be added to make the water really safe to drink. The carbon filters require regular changing because they can quickly become breeding grounds for bacterial growth.

Although distilled water is safe, it is not healthy as this contains no nutrient minerals, which are essential for the drinking purpose. This type of water purification technology is also very slow. Adding to that, the cost of a carbon filter and the result is an unwieldy system of water purification.

Benefits:

- Removes a broad range of contaminants (toxic chemicals, heavy metals, bacteria, viruses, parasites),

- Continuous,

- Does not rely on physical barriers (filters),

- Does not require additional disinfecting process.

Limitations:

- It consumes an enormous amount of energy both in terms of cooling and heating requirements,

- Some contaminants can be carried into the condensate,

- Requires careful maintenance to ensure purity,

- The process is not very effective which are of lower volatility (viz. organics) compared to water.

Adsorption

In this physical process, dissolved contaminants adhere to the porous surface of the solid particles. It is the surface phenomena and the outcome of surface energy. With the material, all the bonding requirements of the constituent atoms of the material are filled with other atoms. However, atoms on the surface of the adsorbent are not wholly surrounded by other adsorbent atoms and physical attractive force results. It can be physisorption (originates from vanderwaals forces) and chemisorption (originates from co-valent forces).

The adsorbent systems are added directly to the water supply or via mixing basin. Adsorbents combine chemical and physical processes to remove the compounds that impart color, taste, and odor to water. In principle, all microporous materials can be used as adsorbents. However, those with well controlled and highly microporous are the most preferred. The porous solids, e.g., activated carbon, silica gels, aluminas, zeolites, etc. contain many cavities or pores with diameters as small as a fraction of a nanometer is useful.

The isotherms are the quantitative interrelation between the adsorbate and adsorbent. The three most well-known isotherms are Freundlich, Langmuir and Linear. The most commonly used for the water contaminants is Freundlich and it is expressed as:

$$\frac{x}{m} = KC_e^{1/n}$$

Where, x is the mass of solute adsorbed, M is the mass of adsorbent, C_e is the equilibrium concentration of solute, and K and n are experimental constants.

Activated Carbon

The most commonly used adsorbent is activated carbon—a substance which is quite similar to common charcoal. Actually, the active carbon is much more efficient because of its high porous character. The high porous character is generated by treating carbon to steam and high temperature (1300 °C) with or without oxygen in the presence of inorganic salts (physical method). The carbon may be of petroleum coke, bituminous coal, lignite, wood products, and coconut/peanut shells. At high temperature, parts of carbon are oxidized in CO_2 and steam. The gases are evacuated and micro fractures and pores are generated in the carbon structure. It dramatically increases the carbon surface area, making a useful material for the removal of contaminants. In some cases, the carbonaceous matter may be treated with a chemical activating agent such as phosphoric acid, zinc chloride and the mixture carbonized at an elevated temperature, followed by the removal of activating agent by water washing (chemical method).

Active carbon uses the physical adsorption process, whereby Vanderwaals attractive forces pull the solute contamination out of the solution and onto its surface. The efficiency of the adsorption depends on the nature of the carbon particle and pore size, surface area, density and hardness as well as the nature of the contaminants (concentration, hydrophobicity, polarity and solubility of the contaminant and contaminant attraction to the carbon surface).

There are two different forms of activated carbon in common use, granular activated carbon (GAC) and powdered activated carbon (PAC). Physically, the two differ as their names suggested by particle size and diameter. The reusability of the carbon is done primarily with the GAC as PAC particles are too small to be reactivated.

Benefits:

- Activated charcoal is effective for trapping carbon-based impurities (volatile organic chemicals), chlorine (including cancer-causing by-product trihalomethanes) as well as colors and odors,

- Very cost effective,

- Long life (high capacity).

Limitations:

- In GAC scheduled filter replacements, it is important to eliminate the possibility of 'channeling' which reduces the contact between the contaminant and the carbon. Therefore, it reduces efficiency, and the accumulation of bacteria in the filter,

- Frequent filter changes often required,

- Can generate carbon fines.

Activated Alumina

Activated alumina consists mainly of aluminum oxide $\left(Al_2O_3\right)$ spherical beads, highly porous and exhibits tremendous surface area. The surface area of activated alumina is in the range 345–$415\,m^2/g$. It does not shrink, swell, soften or disintegrate when immersed in water. It can exist in three forms, viz. activated alumina sorbent, activated alumina desiccant and activated alumina catalyst carrier. The granulated alumina has the internal active surface of the alumina.

In this process, contaminated water is passed through a cartridge or canister of activated alumina. The contaminant adsorbs on the alumina. As the physical adsorption has a particular limit, the cartridge of activated alumina must be replaced periodically.

Benefits:

- Tailoring of activated alumina is possible by varying the activation process and dopant variation,

- Effective in removing As^{5+}, PO_4^{3-}, Cl^-, and F^- from water,

- Removal of Se, Sb, Pb and Bi from the water is also possible.

Limitations:

- The method is not very much capable of reducing levels of other contaminants of health concern. It needs another support.

Zeolite

Zeolites are aluminosilicates with an Si/Al ratio between 1 and infinity. It has a tetrahedral network of silicon and oxygen atoms, and some of the silicon atoms are replaced by aluminum to form alumino-silicates. The adsorptive property of zeolite is considered due to the crystalline nature of the materials. The channels in it are of extended honeycomb and cavities. Zeolites have the surface area $1-20\,m^2\,/\,g$. Synthetic Zeolites are manufactured by hydrothermal processes in a temperature range of 90–100 °C, an autoclave followed by ion exchange with certain cations $\left(Na^+, Li^+, Ca^{2+}, K^+, NH_4^+\right)$. The high cation exchange and molecular sieve properties, such as zeolites, have been widely used as adsorbents. The water softening process is by exchanging Na^+ with the Ca^{2+}/Mg^{2+} in water, as follows:

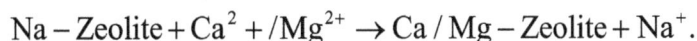

$$Na - Zeolite + Ca^2 + /Mg^{2+} \rightarrow Ca\,/\,Mg - Zeolite + Na^+.$$

Natural zeolites in the waste-water treatment are very useful. Many natural zeolites (e.g., Clinoptilolite, mordenite, phillipsite, chabazite) show selective separation towards NH_4^+ and also for transition metals $\left(e.g., Cu^{2+}, Ag^+, Zn^{2+}, Cd^{2+}, and\ Hg^{2+}\right)$.

Benefits:

- Recharging of zeolite is feasible by exchanging the cations with the initial one and thus reuse is also feasible,

- Removes NH_4^+ and heavy metal removal of inorganic anions (nitrates, phosphates, arsenates, chromates and fluorides) as well as radionuclides (e.g., 137Cs, 90Sr, 60Co, 45Ca, 51Cr, 111mCd, 110mAg) is also possible,

- Removal of organics and other humic substances (including humic, fulvic acid, and humin) and odor is also possible,

- Microorganisms capturing (the large surface area of the zeolites is accessible for adhering microorganisms. This makes selecting a suitable material for biofilter for removal of pathogenic microorganisms),

- By the zeolites, the permeable reactive barriers (PRB) can be prepared in the waste disposal site, so that contaminations could not spread in the ground water.

Limitations:

- As zeolites are used as softener in detergent formulations and insoluble, they lead to increase in sewage sludge mass.

Silica Gel

Silica gel is an amorphous hard glass-like granules or beaded material made of silicon dioxide (SiO_2). Basically, it is a naturally occurring mineral which is purified and processed. Silica gel is a high capacity adsorbent with fine pores on the surface and can be used especially as desiccant, moisture-proof, rust inhibitor as well as catalysis.

Generally, it is formed by two routes: (1) polymerizing silicic acid, and (2) aggregation of particles of colloidal silica. Silicic acid, $Si(OH)_4$, has a strong tendency to polymerize and form a network of siloxane (Si–O–Si), leaving a minimum number of uncondensed Si–O–H groups. The aggregation is by Van der Waals forces or by cations bridging as coagulants. Commercial silica is prepared through the first route by mixing a sodium silicate solution with a mineral acid, such as sulfuric or hydrochloric acid. The reaction produces a concentrated dispersion of finely divided particles of hydrated SiO_2, known as silica hydrosol or silicic acid:

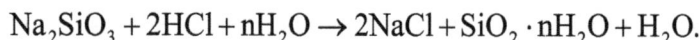

$$Na_2SiO_3 + 2HCl + nH_2O \rightarrow 2NaCl + SiO_2 \cdot nH_2O + H_2O.$$

The hydrosol, on standing, polymerizes into a white jelly-like precipitate, which is silica gel. The resulting gel is washed, dried, and activated. Various silica gels with a wide range of properties, such as surface area, pore volume, and strength, can be made by varying the silica concentration, temperature, pH, and activation temperature. Two common types of silica gel are known as regular-density and low-density silica gels, although they have the same densities (true and bulk). The regular-density gel has a surface area of $750 - 850 \, m^2 / g$ and an average pore diameter of 22–26 A°, whereas the respective values for the low density gel are $300 - 350 \, m^2 / g$ and 100–150 A°).

Because of its large pore volume and mesoporosity, silica gel is used as desiccant. The modified silica gel (modified by the impregnation) with a high-molecular weight quaternary amine (triethyl octadecyl ammonium iodide) has been used for the concentration of heavy metals (Cs, Ag, Hg, Cu, Cd, etc.) for water purification.

Benefits:

- Silica gel is non-toxic, non-corrosive material,

- It has high adsorption capacity because of very high surface area and porosity.

Limitations:

- Preparative aspects needed very precise control,

- Modification is needed to remove the contaminants.

Ion Exchange

The coulombic attractive force between ions and charged functional groups is more commonly classified as ion exchange. It is a typical reversible chemical reaction where an ion from a solution is exchanged for a similarly charged ion attached to an immobile solid particle.

The selectivity coefficient controls the preference for ions of particular resins and is expressed as follows:

$$K_{B^+}^{A^+} = \frac{\{\overline{A}\}\{B^+\}}{\{A^+\}\{\overline{B}\}}$$ for the exchange of $\overline{A}$ in solution for B^+ on the resin:

$$A^+ + \overline{B} \leftrightarrow B^+ + \overline{A}$$

The barred terms indicate location on the resin (solid phase) as opposed to solution phase. The superscript and subscript on the selectivity coefficient indicate the direction of the reaction.

Ion exchange materials are insoluble substances containing loosely held ions, capable of exchanging particular ions within them with ions in a solution that is passed through them. Many natural substances like proteins, cellulose, living cells and soil particles exhibit ion exchange properties, which play an important role in the way the function in nature. Synthetic ion-exchange polymers can be made in two forms, viz. beaded polymer matrix (resins) and membranes.

Ion Exchange Resins

Ion exchange resins are very small polymer matrix (beads), with a diameter of about 0.6–1.0 mm. The ion exchange resins can be manufactured in one of the two physical structures, gel and porous. The gel resins are crosslinked polymers having no porous structure, while porous resins have considerable external and pore surfaces (microporous, mesoporous and macroporous) where ions can attach. The porous polymer matrices contain invisible water inside the pores of the beads, measured as "humidity" or "moisture content". The functional groups (ions) can be attached on the polymer matrix which cannot be removed or displaced.

Based on their functional groups attached on polymer matrix, the ion exchange resins are two types: cation and anion exchange resins, which further subdivided into four categories:

- Strongly acidic (typically, sulfonic acid groups, e.g., sodium polystyrene, sulfonate, etc.,

- Strongly basic (quaternary amino groups, for example, trimethylamonium group,

- Weakly acidic (mostly, carboxylic acid groups),

- Weakly basic (primary, secondary, and/or ternary amino groups, e.g., polyethylene amine).

Cation exchange resins exchange cations like calcium, magnesium, radium, and anion resins, used to remove anions like nitrate, arsenate, arsenite, or chromate from waste solution/water. Regeneration can be possible using sodium chloride. In case of cation resins, sodium ion displaces the cation from the exchange site; whereas in case of anion resins, the chloride ion displaces the anion from the exchange site. Resins can be designed to show a preference for specific ions, so that the process can be easily adapted to a wide range of different contaminants.

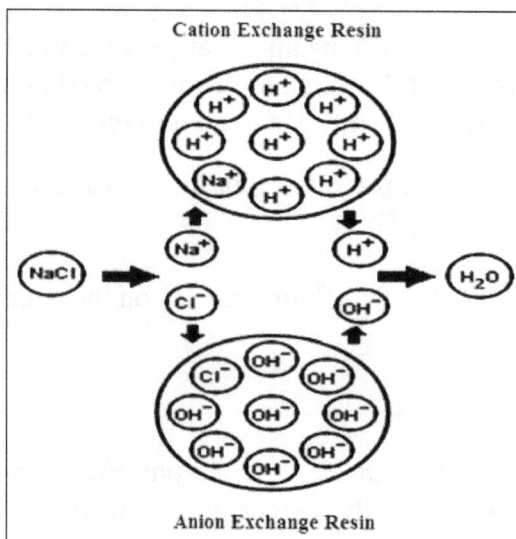

Schematic diagram regarding the behavior cation exchange and anion exchange resin.

The mode of preparation of ion exchange resins is through suspension polymerization technique containing the monomers, cross-linkers and initiators. Various types of polymeric beads like styrene, MMA, MAA, DVB, etc. can be prepared by this technique, varying the ratio of monomers, diluents, the stabilizer, concentration and the agitation rate is dispersed by agitation in a liquid phase, usually water, in which the monomer droplets are polymerized while they are dispersed by continuous agitation also known as pearl polymerization technique.

The most important issue in the practical operation of suspension polymerization is the control of the final particle size distribution. The size of the particles depends on monomer type, monomer purity, interfacial tension, stabilizer concentration, agitation condition in the reactor (degree of agitation) design of reactor/stirrer.

Benefits:

- Simple and low running cost technique,

- Technique is very useful in separating components/contaminations (cations and anions) from dilute solutions/wastes and in water purification, etc.,

- Useful for the recovery of expensive materials from industrial waste (e.g., precious metals),

- Recycling components present in the solutions and/or regenerating chemicals,

- Capability to handle hazardous wastes,

- Simple regeneration process and well-maintained resins last for many years.

Limitations:

- Limitation on the concentration in the effluent to be treated,

- Ion exchange resin-treated water contains sodium, which cannot be recommended for the diet requiring low sodium intake,

- Generation of waste (sodium wastewater) as a result of ion exchange regeneration,

- Ion exchange resins do not remove organic compounds or biological contaminants,

- If resin is not sanitized or regenerated regularly, bacterial colonies proliferate on resin surfaces and can contaminate drinking water.

The ion exchange membranes are discussed in the following part.

Apart from the above, interests are growing to develop different low cost adsorbents. For this purpose, numerous agro-waste biomaterials are found suitable, viz. rice-husk, soyabean hulls, coconut shells, rice straw, sugarcane bagasse, tea leaves, petiolar felt-sheath of palm trees, etc.. These are useful for removal of heavy metal ions $\left(Pb^{2+},\ Ni^{2+},\ Cd^{2+},\ Zn^{2+},\ etc. \right)$ in low concentrations. Biosorption is a rapid phenomenon of passive metal uptake sequestration of non-growing biomass. Biomass of Aspergillusniger, Penicillum Chrysogenoum, Rhizopusnigricans, Ascophyllumnodosum, Sargassumnatans, Chlorella fusca, Oscillatoriaanguistissima, Bacillus firmus and Streptomyces sp. has also the potential to sequester metal ions by forming metal complexes from solution and obviates the necessity to maintain special growth-supporting conditions.

Membrane Water Treatment

Membrane technology is one of the innovative ideas of water treatment. Over here, a semipermeable membrane is used for the removal of water impurities. There are two types of membrane water treatment technologies, namely pressure-driven (e.g., reverse osmosis) and electrically driven (electro-membrane).

Reverse Osmosis

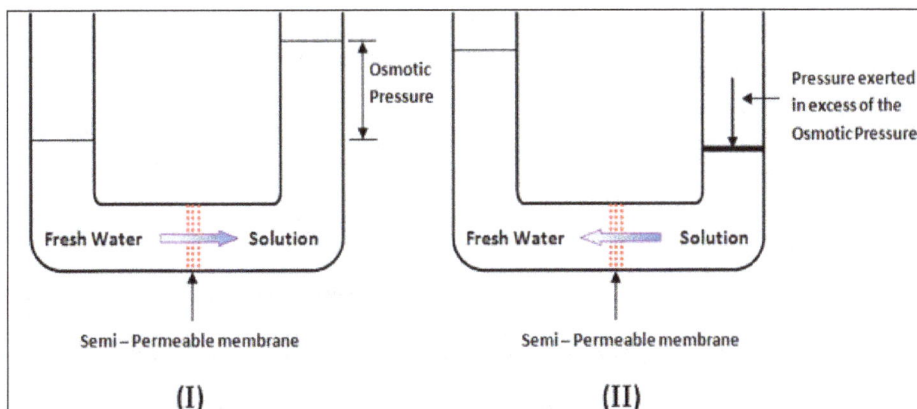

Schematic diagram of osmosis and reverse osmosis.

The two processes (viz. osmosis and reverse osmosis) are the regulator of life. Though they are termed as concentration and pressure driven simultaneously, both are controlled by thermodynamic function, i.e., 'chemical potential' of the systems. It is essentially a driving force expressed as a change in the free energy of the system as a result of the change in the composition of the system. Though literally the two signify just the opposite process, thermodynamically they are similar. Under isothermal operating condition, the tendency for material transport is always in the direction of lower chemical potential for both the processes. In osmosis, the flow is occurring solvent to solution side through a semipermeable membrane, whereas in reverse osmosis the flow is a solution for solvent. In both cases, only solvent molecules migrate from one side to another. The schematic diagram of osmosis and reverse osmosis is presented in figure.

The main two characteristics of a membrane process are flux and rejection. If an RO membrane is considered as permeating water only, the water and solute flux can be written as:

$$J_w = A(\Delta P - \Delta \Pi),$$
$$J_s = B\Delta c_s,$$

Where, A is the permeability coefficient, and ΔP and $\Delta \Pi$ are the hydraulic pressure and osmotic pressure difference across the membrane and B is the solute permeability coefficient and ΔC_s is the solute concentration difference across the membrane.

The microfiltration and ultrafiltration membranes have a pore size in the range of >10 and 1–100 nm, respectively, whereas in the case of nanofiltration and reverse osmosis membranes are in the range of ~1 and <1 nm. Size selective separation operates in case of micro and ultrafiltration, whereas the size and charge selective separation operate in the latter two.

The membranes are generally based on natural and synthetic polymers (cellulose acetate, cellulose triacetate, polysulfone, polyamide, etc.). The most popular RO membrane is thin film composite membranes (i.e., polyamide layer on asymmetric polysulfone). The polyamide layer is formed by interfacial polymerization of diamine and acyl halide and shows the charge holding capacity in it.

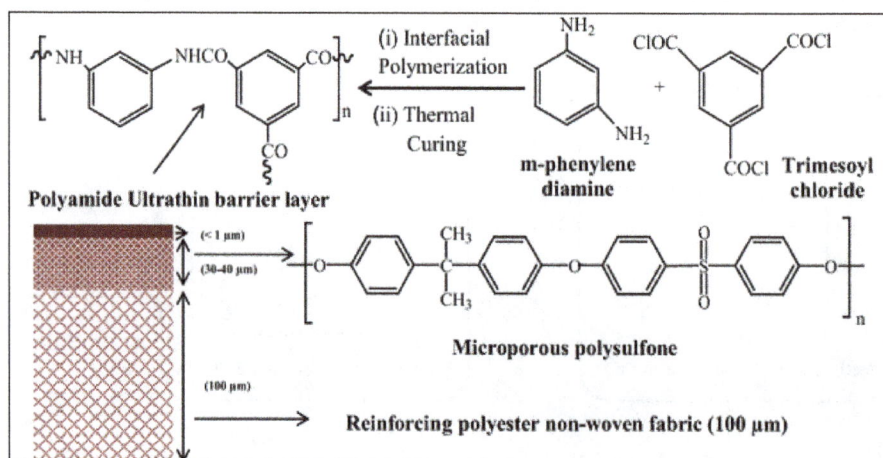

Schematic diagram of polyamide thin film composite membrane.

Reverse osmosis (RO) is one of the most effective types of water treatment and widely used water purification processes in the world. It is usually used for home water treatment to remove salts, chemical toxins, organic contaminants, dyes, pesticides and microbes. In reverse osmosis, the raw water is forced (with pressure) through a dense membrane filter that prevents passing of impurities.

Benefits:

- No phase changes and thus requirement of low energy,

- Eco-friendly as they do not produce or use any harmful chemicals; compactness and space requirements are less compared to distillation, and can be designed according to the requirement,

- Ability to remove almost all kinds of contaminates like Cl^-, NO_3^-, F^-, SO_4^-, Pb^{2+}, Na^+, K^+, Mg^{2+}, organics as well as microorganisms,

- No alteration in the taste and smell of water and effective removal of microbes and toxins.

Limitations:

- The purified water obtained after reverse osmosis treatment is devoid of useful minerals,

- Membrane may become clogged after prolonged use and, hence, requires periodical replacement of the membrane.

Electrodialysis Membrane Treatment

Electrodialysis (ED) is electric potential-driven membrane-based separation process. The basic principle of the membrane separation is similar to ion exchange reactions. The charged groups are attached to the polymer backbone of the membrane material and it is obvious that the fixed charge groups partially or completely exclude ions of the same charge from the membrane, i.e., an anionic membrane with fixed positive groups excludes positive ions, but is freely permeable to negatively charged ions whereas cationic membrane with fixed negative groups excludes negative ions but is freely permeable to positively charged ions.

Since the membrane is of ion selective, it separates or rejects opposite charge ions, useful in removal, or separation of electrolytes. The schematic diagram is presented in figure.

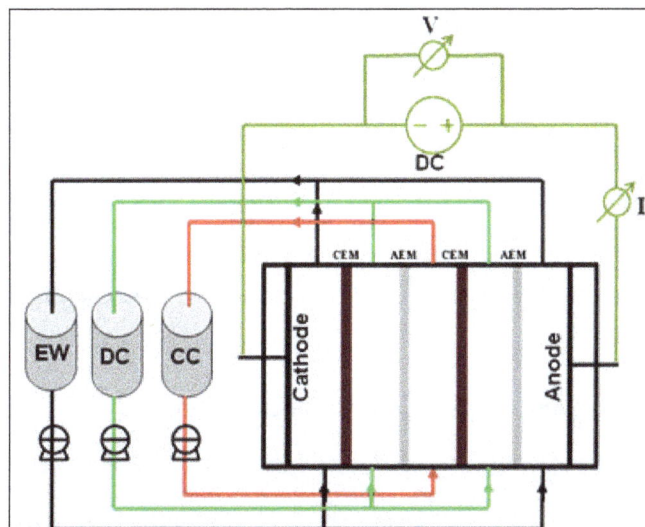

Schematic diagram of electrodialysis. DC diluted chamber, EW electrode wash, CC concentrated chamber, CEM cation exchange membrane, AEM anion exchange membrane.

The ion transportation depends on the current efficiency in the particular system. Generally, the current efficiencies >80 % are required in commercial stacks to minimize energy operating costs. The low current efficiencies result in water splitting in the dilute or concentrate streams, shunt currents between the electrodes, or back-diffusion of ions from the concentrate to the dilute. The current efficiency is calculated according to the following equation.

$$\xi = zFQ_f (C_{inlet} - C_{oulet}) / NI$$

where ξ is the current utilization efficiency, z is the charge of the ion, F is the Faraday constant 96485 Amp-s/mol, Q_f is the dilute flow, L/s C_{inlet} is the dilute ED cell inlet concentration, mol/L, C_{outlet} is the ED cell outlet concentration, mol/L, N is the number of cell pairs, and I is the current, Amps.

Apart from their chemical structure (cation and anion), the commercial ion-exchange membranes can be divided, according to their structure and preparation procedure, into two major categories, homogeneous and heterogeneous and depending on the degree of heterogeneity of the ion-exchange membranes, these can be further classified into different types: mono polar (cation/anion) ion-exchange, amphoteric ion-exchange, bipolar ion-exchange, inter-polymer membranes.

The ion-exchange membranes are very similar to normal ion-exchange resins in terms of chemical structure as well as of high selectivity and low resistivity. The difference between membranes and resins arises largely from the mechanical requirement of the membrane process. Thus, it is generally not possible to use sheets of material that have been prepared in the same way as a bead resin. However, the most common solution to this problem is the preparation of membrane with a backing of a stable reinforcing material that gives the necessary strength and dimensional stability. The preparation method of ion-exchange membranes can be summarized in three different steps, viz. Polymerization or polycondensation of monomers; at least one of them must contain a moiety that either is or can be made anionic or cationic groups, respectively, introduction of anionic or cationic moieties into a pre-formed solid film such as styrene-DVB-based membrane, and introduction of anionic or cationic moieties into a polymer, such as Polysulfone, followed by dissolving the polymer and casting it onto a film.

Benefits:

- Non-pollution, safety and reliability,

- Completely eliminated the chemicals for regeneration,

- Effective for complete removal of dissolved ionic particles (cation and anions), heavy metals, etc.,

- Ability to treat feed water with higher SDI, TOC and silica concentrations.

Limitations:

- Removal of low-molecular weight ionic contaminations,

- Non-charged, higher molecular weight, and less mobile ionic species cannot be significantly removed by the process,

- Large membrane areas are required to satisfy capacity requirements for low concentration (and sparingly conductive) feed solutions.

Catalytic Processes

Catalytic processes are typically achieved by the following three methods: hydrogenation of nitrate, photocatalytic and electrocatalytic.

Hydrogenation of Nitrate

The hydrogenation via catalytic method is one of the promising techniques for removal of nitrate from water. It needs very active catalysts because the reaction is performed preferably at an ambient/low temperature. The reaction scheme shows that nitrate is reduced to the desired products involving NO_2^-, NO, N_2O and N_2. The undesired byproduct NH_4^+ is also formed by a side reaction due to over hydrogenation. Supported bimetallic catalyst (viz. Pd/Cu, Pd/In, and Pd/Sn) has emerged as efficient catalysts for nitrate hydrogenation. Apart from Pd, the other metals (e.g., Cu, In, Sn, Co) serve as the role of promoter for the first reduction step to convert NO_3^- into NO_2^-. It is seen in the schematic reaction, below that N_2 and ammonium $\left(NH_4^+\right)$ are the stable end products of the catalytic reduction process. N_2 is not harmful, but the second one is considered a hazardous aquatic pollutant. That is why target is to convert NO_3^- into N_2 as an end product.

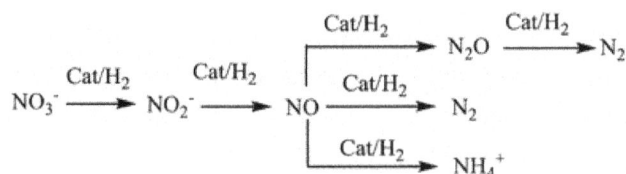

$$NO_3^- \xrightarrow{Cat/H_2} NO_2^- \xrightarrow{Cat/H_2} NO \begin{cases} \xrightarrow{Cat/H_2} N_2O \xrightarrow{Cat/H_2} N_2 \\ \xrightarrow{Cat/H_2} N_2 \\ \xrightarrow{Cat/H_2} NH_4^+ \end{cases}$$

Benefits:

- The method can be of single operation mode,

- Selectivity of catalyst can counter the formation of ammonia ions,

- Addition of other chemicals can be avoided.

Limitations:

- Increase in pH in the reaction medium forms ammonia in dissolved condition, which is more harmful than nitrate.

Photocatalytic Method

The method is based on the acceleration of photodegradation of organic pollutants, pathogens, green algae, and substances in the presence of catalyst. In response to UV light, when they excited charge separation followed by scavenging e–s and holes by surface adsorbed species. The heterogeneous photocatalysts employing semiconductor catalysts $\left(TiO_2, ZnO, Fe_2O_3,\right)$ have shown their efficiency in degrading a wide range of pollutants in water. Metal oxides are more suitable, since they are more resistant to poisoning and deactivation.

Upon UV-irradiation, photocatalytic reactions are initiated by the absorption of illumination with photo-energy equal to or greater than the band gap of the semiconductor. It results in electron–hole $\left(e-/h^+\right)$ pairs as shown in figure. Thus, it participates in the redox reaction with the adsorbed pollutant species in water. Apart from the reaction, the semiconductor also oxidizes water to produce OH•, a powerful oxidant, which rapidly reacts with the pollutants in the water.

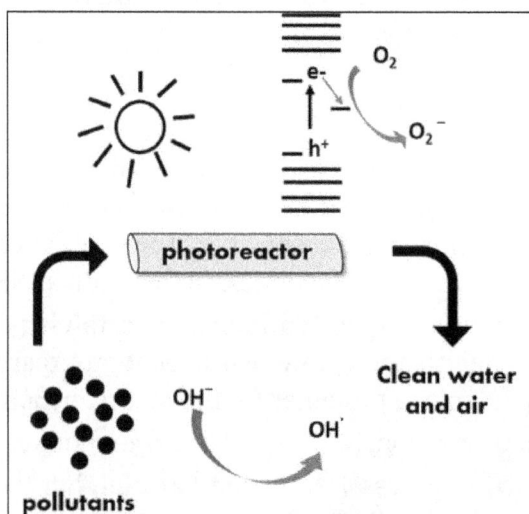

Schematic diagram of the photocatalytic arrangement.

To improve the catalytic activity using visible light, various approaches are also developed, viz. addition of dopants, stoichiometry of catalytic metal oxides and mixed metal oxides, particle size and shape. TiO_2 doped with nitrogen showed excellent photo catalytic activities compared to unmodified TiO_2 nanoparticles in both degradation of chemicals and bactericidal reaction.

Benefits:

- Reusability of the catalyst as it is unchanged during the process,

- Reactions can occur in ambient condition as well as no consumable chemicals are required,

- Operational process is simple,

- It is good enough to treat low concentration of pollutants.

Limitations:

- Post-separation of the semiconductor catalysts after water treatment is important and failing results in catalyst poisoning.

- The catalysts with their fine particle size and large surface area to volume ratio create a strong catalyst agglomeration tendency during the operation.

Electrocatalytic Oxidation

In the electrocatalysis, the oxidation occurs through surface mediator on the anodic surface. The rate of oxidation depends on temperature, pH and diffusion rate of generating oxidants in indirect electrolysis. This is somewhat different from electrolysis where direct oxidation of pollutants takes place and rate of oxidation depends on electrode activity, pollutants diffusion rate and current density.

The electrocatalysis through metal oxide (MO_x) electrode can be shown as follows:

$$MO_x + H_2O \rightarrow MO_x(\cdot OH).$$

In the presence of organics (R) present in waste water, the physiosorbed active oxygen $(\cdot OH)$ involves in complete combustion of organics (1) and chemisorbed active oxygen in the form of MO_{x+1} (2) does the selective oxidation:

$$R + MO_x (\cdot OH) \rightarrow CO_2 + H^+ + e + MO_{x+1},$$

$$R + MO_{x+1} \rightarrow RO + MO_x.$$

The key role in the electrocatalytic process is electrocatalytic material. Ru/Pb/Sn oxide and Pb/PbO_2 coated with Ti is used in the dye oxidation. Pt, TiO_2, IrO_2, PbO_2, several Ti-based alloys and boron-doped diamond (BDD) electrodes are employed for the removal of effluents containing various organics, viz. phenols, pharmaceuticals, alcohols, carboxylic acids, anionic surfactants and pesticides. $Pt(acac)_2$ onto ruthenium nanoparticles is used for the removal of formic acid.

Electrocatalytic reduction is largely used for NO_3^- removal. In this regard, the development of electrodes (viz. Ti–Rh, Ti/IrO_2–Pt, PPy–Graphite, Carbon cloth–Rh, Pd–Sn activated carbon fiber is interesting direction.

Benefits:

- High pollutant degradation, easy control and low cost,
- It can be easily controlled by putting on/off the power,
- Environmentally compatible since there is little or no need for additional chemicals,
- It has the potential to eliminate different types of pollutants as well as bulk volume,
- It operates at low temperature and pressure compared to non-electrochemical methods; thus, the volatilization and discharge of un-reacted wastes can also be avoided.

Limitations:

- High operating cost due to the high energy consumption during operation,
- Electrode fouling may also occur on the surface of the electrodes,
- It needs, conducting nature of the effluent. Sometimes the addition of an electrolyte may be necessary,
- The use of metal ions resulted in an effluent with a higher toxicity than that of the initial effluent. Thus, this approach requires a separation step to recover the metallic species.

Bioremediation

Phytoremediation

It signifies the removal of pollutants from the environment by the use of plants. The technology involves different mechanisms, viz. phytoextraction, rhizofiltration, phytostabilization, phytotransformation/phytodegradation. Phytoextraction involves metal accumulation into the harvestable parts of the roots and the above ground shoot. Rhizofiltration indicates the absorption, precipitation and concentration of toxic metals from polluted effluents. Phytostabilization is a process in which mobility of heavy metals is reduced through the use of tolerant plants, whereas phytotransformation/phytodegradation is the process in which contaminants can be eliminated via phytodegradation or phytotransformation by plant enzymes or enzyme co-factors.

The history of the particular study, including the uptake of toxic metals Hg, As, and other metals, begins in the 70's and other metals. In this regard, macrophytes water hyacinth (Eichhorniacrassipes); pennywort (Hydrocotyle umbellate L); duckweeds (Lemna minor L.) and water velvet (Azollapinnata) are considered the biological filters and play the important role in the maintenance of the aquatic ecosystem. The floating plants Lemna minor, Eichhorniacrassipes and Pistiastratiotesand Salviniaherzogii show good potential in accumulating the metals directly from industrial effluents.

Benefits:

- Cost effective,
- Eco-friendly.

Limitations:

- Seasonal growth of the plants,
- Biomass disposal.

Vegetated Filter Strips

The filter strips are meant as land areas of either planted or indigenous vegetation, situated between a potential, pollutant-source area and a surface-water body that receives runoff. Vegetated filter strips (viz. grassed filter strips, filter strips, and grassed filters) are vegetated surfaces that are designed to treat sheet flow from adjacent surfaces. The run-off usually carries sediment, organics, plant nutrients and pesticides.

The trapped plant nutrients and pesticides may be easily degraded or transformed by biological and chemical processes. Cole et al. report the removal of chloropyrifos (62–99 %), dicamba (90–100 %), 2,4D (89–98 %), and mecoprop (89–95 %) using Bermuda grass buffer. On the other hand, atrazine (98 %) and pyrethroid (100 %) removal is possible using vegetated drainage ditch.

Benefits:

- Trap sediments,
- Capture nutrients both through plant uptake and adsorption of soil particles,
- Promote transformation and degradation of pollutants into less toxic forms,
- Removal of pathogens is possible.

Limitations:

- The design is important,
- Proper vegetation is necessary.

Biologically Active Carbon Filtration

Biologically active carbon is another prospective process with this bioremediation technique. The process utilizes granulated activate carbon (GAC) as its water filtration. The microbial (bacterial) colonization is possible over the GAC media particles form 'biofilm'. Actually, it is described as a 'porous tangled mass of slime matrix. It consists of microbial cells, either immobilized on the surface of the GAC (substratum) or embedded in an extracellular microbial organic polymer matrix. Bacteria and fungi cells in the biofilms secrete extracellular polymeric substances to form a cohesive, stable matrix in which cells are held in dense agglomeration. The extracellular matrix is composed of polysaccharides, proteins, nucleic acids and lipids. The activity of the biofilm relates to the physiological modifications associated with the promotion of certain genes, or changes the bacteria cell surrounding to increase the local concentration of nutrients, oxygen and enzymes or limit the invasion of toxic or inhibiting substances.

Most of the dissolved organic chemical removal occurs through physical adsorption in the GAC media. Apart from the adsorption, biodegradation can also operate.

Benefits:

- It can avoid chemical disinfection water treatment processes,
- Because of the microbial biodegradation of organic substrates on the GAC media, the service life can be extended,
- Bacterial regrowth is less possible,
- Eliminates the need for coagulant in source filtration processes.

Limitations:

- The control of the growth of the process is necessary.

Magnetic Separation

In the magnetic separation process, the high-gradient magnetic separation (HGMS) is a commonly used process. In this case, device comprising bed of magnetically susceptible wires placed inside an electromagnet is used. There are various influencing factors, viz. nature of impurities, concentration, size, magnetic susceptibility, spacing design, and intensity of magnetic field and its orientation, magnetic field strength.

Generally, there are three categories of separators based on magnet type, viz. permanent magnet, electromagnet and superconducting magnet. The permanent magnet (ferro-magnets of iron-based, nickel, cobalt or rare earth element) is having magnetic fields of less than 1 T, though trend is to improve the magnetic field strength by the development of materials and shape design parameters. The electromagnetic-based device consists of a solenoid of electrically conducting wires which can generate a magnetic field of 2–4 T within their cavity on passage of electric current. The third category of magnetic separators generates the highest intensity magnetic field from 2 to 10 T.

The magnetically assisted water purification can be primarily classified into the following type depending on the difference in adoption of physical processes, viz. direct purification, seeding and separation by magnetic flocculant, and magnetic sorbent application in organic and inorganic contaminants including radionuclides. In the direct purification technique, there is no carrier magnetic component utilized for the separation. The basic properties of ions or solid response to the magnetic field are utilized for purification. In this method, the anti-scaling technique is most commonly practiced. In the area of anti-scale magnetic treatment, the most common constituents of scale are $CaCO_3$, $CaSO_4 \cdot 2H_2O$ and silica, $BaSO_4$, $SrSO_4$, $Ca_3(PO_4)_2$ and ferric and aluminum hydroxides. In the magnetic flocculant separation, coagulant cation [viz. Fe(III)] forms an insoluble precipitate under applied magnetic field. It is an effective means of lowering significantly both the oil and suspended solids of water effluent streams. Ions (polymerise as polyhydroxycomplexes, or nitroso-hydroxy, or hydroxy-carbonato or halogenohydroxo-carbonato complexes), which are difficult to coagulate magnetic sorbents, are utilized for waste water purification.

Benefits:

- Useful for the separation of pollutants,
- Magnetic pre-treatment improvises purification RO membrane filters,
- Calcium carbonate scale formation in heat exchanger can be reduced,
- Promotes the homogeneous precipitation of calcium carbonate scales.

Limitations:

- Not fully sufficient.

Hybrid Technologies

In true sense, no technologies independently counter all the problems. The development of technology is a dynamic process that moves forward slowly and recommendations are made based on the best science available at that time. However, with new research and new results, the flaws of existing technologies may be removed. That is why the concept of the combinations of technologies or in other sense hybrid technologies has come. Scientists and technologists have orchestrated according to the requirement. Let us discuss with the synergistic RO technology first. In the RO technology, feed pretreatment is vital for RO to avoid problems, i.e., fouling, damaging the membranes. Conventional pretreatment steps include chemicals addition, i.e., acid, coagulant/flocculant, disinfection. Coagulation and flocculation (coagulants–flocculants) are dealt in water treatment process. Chlorine treatment is treated as disinfection process and commonly employed. But chlorination shortens the stability of the membrane and, thus, dechlorination treatment (viz. sodium bi sulfite) is required. In media filtration, water is treated by passing through granular media like pumice, anthracite, gravels, etc. that can be used in combination. Cartridge filter (made up of papers, woven wire, cloth) is used as the last pretreatment step to retain particles in the size range 1–10µm. To check the quality, 'Silt Intensity Index' or SDI parameter is important. Actually, SDI considers the ratio of two flow measurements, one at the beginning, and the other at the end by passing feed water through a 0.45 µm filter paper in dead end mode at constant pressure.

Similarly, the pretreatment step coagulation is coupled with ion exchange treatment of water. The coupled electrodeionization technology based on electrodialysis and ion exchange results in a process which effectively deionizes water, while the ion exchange membranes are continuously regenerated by the electric current in the unit. This electrochemical regeneration replaces the chemical regeneration of conventional ion exchange systems. Recently, hybrid technologies like ED-RO or ED-RO and distillation have been developed for the water purification and these processes offer many advantages over the traditional technologies. The ED-RO technologies desalinate the brackish water with high recoveries along with zero discharge and reduced energy consumption. ED-RO is a high recovery system since RO concentrates can be recycled through the ED system to reduce the feed flow rate, pre-treatment cost and the reduced amount of effluent. Thus, coupling of the technologies/processes offers a solution to an increasing important issue in water treatment as well as for water conservation. To get better results, UV is typically used as a final purification stage in terms of removing contaminants bacteria and viruses.

Industrial Water Treatment

Before raw sewage can be safely released back into the environment, it needs to be treated correctly in a water treatment plant. In a water treatment plant, sewage goes through a number of chambers and chemical processes to reduce the amount and toxicity of the waste:

- The sewage first goes through a primary phase. This is where some of the suspended, solid particles and inorganic material is removed by the use of filters.

- The secondary phase of the treatment involves the reduction of organic, this is done with the use of biological filters and processes that naturally degrade the organic waste material.

- The final stage of treatment is the tertiary phase; this stage must be done before the water can be reused. Almost all solid particles are removed from the water and chemical additives are supplied to get rid of any left-over impurities.

Denitrification

Denitrification is an ecological approach that can be used to prevent the leaching of nitrates in soil, this in turn stops any ground water from being contaminated with nutrients:

- Fertilisers contain nitrogen, and are often applied to crops by farmers to help plant growth and increase the yield.

- Bacteria in the soil convert the nitrogen in the fertilizer to nitrates, making it easier for the plants to absorb.

- Immobilization is a process where the nitrates become part of the soil organic matter.

- When oxygen levels are low, another form of bacteria then turns the nitrates into gases such as nitrogen, nitrous oxide and nitrogen dioxide.

- The conversion of these nitrates into gas is called denitrification. This prevents nitrates from leaching into the soil and contaminating groundwater.

Septic Tanks and Sewage Treatment

Septic tanks treat sewage at the place where it is located, rather than transporting the waste through a treatment plant or sewage system. Septic tanks are usually used to treat sewage from an individual building:

- Untreated sewage from a property flows into the septic tank and the solids are separated from the liquid.

- Solid material is separated depending on their density. Heavier particles settle at the bottom of the tank whereas lighter particles, such as soap scum, will form a layer at the top of the tank.

- Biological processes are used to help degrade the solid materials.

- The liquid then flows out of the tank into a land drainage system and the remaining solids are filtered out.

Ozone Wastewater Treatment

Ozone wastewater treatment is a method that is increasing in popularity. An ozone generator is used to break down pollutants in the water source:

- The generators convert oxygen into ozone by using ultraviolet radiation or by an electric discharge field.

- Ozone is a very reactive gas that can oxidise bacteria, moulds, organic material and other pollutants found in water.

- Using ozone to treat wastewater has many benefits:
  - Kills bacteria effectively.
  - Oxidises substances such as iron and sulphur so that they can be filtered out of the solution.
  - There are no nasty odours or residues produced from the treatment.
  - Ozone converts back into oxygen quickly, and leaves no trace once it has been used.

- The disadvantages of using ozone as a treatment for wastewater are:
  - The treatment requires energy in the form of electricity; this can cost money and cannot work when the power is lost.
  - The treatment cannot remove dissolved minerals and salts.
  - Ozone treatment can sometimes produce by-products such as bromate that can harm human health if they are not controlled.

Groundwater Remediation

Groundwater remediation is the process that is used to treat polluted groundwater by removing the pollutants or converting them into harmless products. Groundwater is water present below the ground surface that saturates the pore space in the subsurface. Globally, between 25 per cent and 40 per cent of the world's drinking water is drawn from boreholes and dug wells. Groundwater is also used by farmers to irrigate crops and by industries to produce everyday goods. Most groundwater is clean, but groundwater can become polluted, or contaminated as a result of human activities or as a result of natural conditions.

The many and diverse activities of humans produce innumerable waste materials and by-products. Historically, the disposal of such waste have not been subject to many regulatory controls. Consequently, waste materials have often been disposed of or stored on land surfaces where they percolate into the underlying groundwater. As a result, the contaminated groundwater is unsuitable for use.

Current practices can still impact groundwater, such as the over application of fertilizer or pesticides, spills from industrial operations, infiltration from urban runoff, and leaking from landfills. Using contaminated groundwater causes hazards to public health through poisoning or the spread of disease, and the practice of groundwater remediation has been developed to address these issues. Contaminants found in groundwater cover a broad range of physical, inorganic chemical, organic chemical, bacteriological, and radioactive parameters. Pollutants and contaminants can be removed from groundwater by applying various techniques, thereby bringing the water to a standard that is commensurate with various intended uses.

Techniques

Ground water remediation techniques span biological, chemical, and physical treatment technologies. Most ground water treatment techniques utilize a combination of technologies. Some of the biological treatment techniques include bioaugmentation, bioventing, biosparging, bioslurping, and phytoremediation. Some chemical treatment techniques include ozone and oxygen gas injection, chemical precipitation, membrane separation, ion exchange, carbon absorption, aqueous chemical oxidation, and surfactant enhanced recovery. Some chemical techniques may be implemented using nanomaterials. Physical treatment techniques include, but are not limited to, pump and treat, air sparging, and dual phase extraction.

Biological Treatment Technologies

Bioaugmentation

If a treatability study shows no degradation (or an extended lab period before significant degradation is achieved) in contamination contained in the groundwater, then inoculation with strains known to be capable of degrading the contaminants may be helpful. This process increases the reactive enzyme concentration within the bioremediation system and subsequently may increase contaminant degradation rates over the nonaugmented rates, at least initially after inoculation.

Bioventing

Bioventing is an in situ remediation technology that uses microorganisms to biodegrade organic constituents in the groundwater system. Bioventing enhances the activity of indigenous bacteria and archaea and stimulates the natural in situ biodegradation of hydrocarbons by inducing air or oxygen flow into the unsaturated zone and, if necessary, by adding nutrients. During bioventing, oxygen may be supplied through direct air injection into residual contamination in soil. Bioventing primarily assists in the degradation of adsorbed fuel residuals, but also assists in the degradation of volatile organic compounds (VOCs) as vapors move slowly through biologically active soil.

Biosparging

Biosparging is an in situ remediation technology that uses indigenous microorganisms to biodegrade organic constituents in the saturated zone. In biosparging, air (or oxygen) and nutrients (if needed) are injected into the saturated zone to increase the biological activity of the indigenous microorganisms. Biosparging can be used to reduce concentrations of petroleum constituents that are dissolved in groundwater, adsorbed to soil below the water table, and within the capillary fringe.

Bioslurping

Bioslurping combines elements of bioventing and vacuum-enhanced pumping of free-product that is lighter than water (light non-aqueous phase liquid or LNAPL) to recover free-product from the groundwater and soil, and to bioremediate soils. The bioslurper system uses a "slurp" tube that extends into the free-product layer. Much like a straw in a glass draws liquid, the pump draws liquid (including free-product) and soil gas up the tube in the same process stream. Pumping lifts LNAPLs, such as oil, off the top of the water table and from the capillary fringe (i.e., an area just above the saturated zone, where water is held in place by capillary forces). The LNAPL is brought to the surface, where it is separated from water and air. The biological processes in the term "bioslurping" refer to aerobic biological degradation of the hydrocarbons when air is introduced into the unsaturated zone contaminated soil.

Phytoremediation

In the phytoremediation process certain plants and trees are planted, whose roots absorb contaminants from ground water over time. This process can be carried out in areas where the roots can tap the ground water. Few examples of plants that are used in this process are Chinese Ladder fern Pteris vittata, also known as the brake fern, is a highly efficient accumulator of arsenic. Genetically altered cottonwood trees are good absorbers of mercury and transgenic Indian mustard plants soak up selenium well.

Permeable Reactive Barriers

Certain types of permeable reactive barriers utilize biological organisms in order to remediate groundwater.

Chemical Treatment Technologies

Chemical Precipitation

Chemical precipitation is commonly used in wastewater treatment to remove hardness and heavy metals. In general, the process involves addition of agent to an aqueous waste stream in a stirred reaction vessel, either batchwise or with steady flow. Most metals can be converted to insoluble compounds by chemical reactions between the agent and the dissolved metal ions. The insoluble compounds (precipitates) are removed by settling and filtering.

Ion Exchange

Ion exchange for ground water remediation is virtually always carried out by passing the water downward under pressure through a fixed bed of granular medium (either cation exchange media and anion exchange media) or spherical beads. Cations are displaced by certain cations from the solutions and ions are displaced by certain anions from the solution. Ion exchange media most often used for remediation are zeolites (both natural and synthetic) and synthetic resins.

Carbon Absorption

The most common activated carbon used for remediation is derived from bituminous coal. Activated carbon absorbs volatile organic compounds from ground water by chemically binding them to the carbon atoms.

Chemical Oxidation

In this process, called In Situ Chemical Oxidation or ISCO, chemical oxidants are delivered in the subsurface to destroy (converted to water and carbon dioxide or to nontoxic substances) the organics molecules. The oxidants are introduced as either liquids or gasses. Oxidants include air or oxygen, ozone, and certain liquid chemicals such as hydrogen peroxide, permanganate and persulfate. Ozone and oxygen gas can be generated on site from air and electricity and directly injected into soil and groundwater contamination. The process has the potential to oxidize and/or enhance naturally occurring aerobic degradation. Chemical oxidation has proven to be an effective technique for dense non-aqueous phase liquid or DNAPL when it is present.

Surfactant Enhanced Recovery

Surfactant enhanced recovery increases the mobility and solubility of the contaminants absorbed to the saturated soil matrix or present as dense non-aqueous phase liquid. Surfactant-enhanced recovery injects surfactants (surface-active agents that are primary ingredient in soap and detergent) into contaminated groundwater. A typical system uses an extraction pump to remove groundwater downstream from the injection point. The extracted groundwater is treated aboveground to separate the injected surfactants from the contaminants and groundwater. Once the surfactants have separated from the groundwater they are re-used. The surfactants used are non-toxic, food-grade, and biodegradable. Surfactant enhanced recovery is used most often when the groundwater is contaminated by dense non-aqueous phase liquids (DNAPLs). These dense compounds, such as trichloroethylene (TCE), sink in groundwater because they have a higher density than water. They then act as a continuous source for contaminant plumes that can stretch for miles within an aquifer. These compounds may biodegrade very slowly. They are commonly found in the vicinity of the original spill or leak where capillary forces have trapped them.

Permeable Reactive Barriers

Some permeable reactive barriers utilize chemical processes to achieve groundwater remediation.

Physical Treatment Technologies

Pump and Treat

Pump and treat is one of the most widely used ground water remediation technologies. In this process ground water is pumped to the surface and is coupled with either biological or chemical treatments to remove the impurities.

Air Sparging

Air sparging is the process of blowing air directly into the ground water. As the bubbles rise, the contaminants are removed from the groundwater by physical contact with the air (i.e., stripping) and are carried up into the unsaturated zone (i.e., soil). As the contaminants move into the soil, a soil vapor extraction system is usually used to remove vapors.

Dual Phase Vacuum Extraction

Dual-phase vacuum extraction (DPVE), also known as multi-phase extraction, is a technology that uses a high-vacuum system to remove both contaminated groundwater and soil vapor. In DPVE systems, a high-vacuum extraction well is installed with its screened section in the zone of contaminated soils and groundwater. Fluid/vapor extraction systems depress the water table and water flows faster to the extraction well. DPVE removes contaminants from above and below the water table. As the water table around the well is lowered from pumping, unsaturated soil is exposed. This area, called the capillary fringe, is often highly contaminated, as it holds undissolved chemicals, chemicals that are lighter than water, and vapors that have escaped from the dissolved groundwater below. Contaminants in the newly exposed zone can be removed by vapor extraction. Once above ground, the extracted vapors and liquid-phase organics and groundwater are separated and treated. Use of dual-phase vacuum extraction with these technologies can shorten the cleanup time at a site, because the capillary fringe is often the most contaminated area.

Monitoring-well Oil Skimming

Monitoring-wells are often drilled for the purpose of collecting ground water samples for analysis. These wells, which are usually six inches or less in diameter, can also be used to remove hydrocarbons from the contaminant plume within a groundwater aquifer by using a belt-style oil skimmer. Belt oil skimmers, which are simple in design, are commonly used to remove oil and other floating hydrocarbon contaminants from industrial water systems.

A monitoring-well oil skimmer remediates various oils, ranging from light fuel oils such as petrol, light diesel or kerosene to heavy products such as No. 6 oil, creosote and coal tar. It consists of a continuously moving belt that runs on a pulley system driven by an electric motor. The belt material has a strong affinity for hydrocarbon liquids and for shedding water. The belt, which can have a vertical drop of 100+ feet, is lowered into the monitoring well past the LNAPL/water interface. As the belt moves through this interface, it picks up liquid hydrocarbon contaminant which is removed and collected at ground level as the belt passes through a wiper mechanism. To the extent that DNAPL hydrocarbons settle at the bottom of a monitoring well, and the lower pulley of the belt skimmer reaches them, these contaminants can also be removed by a monitoring-well oil skimmer.

Typically, belt skimmers remove very little water with the contaminant, so simple weir-type separators can be used to collect any remaining hydrocarbon liquid, which often makes the water suitable for its return to the aquifer. Because the small electric motor uses little electricity, it can be powered from solar panels or a wind turbine, making the system self-sufficient and eliminating the cost of running electricity to a remote location.

Sewage Treatment

The purpose of tertiary treatment is to provide a final treatment stage to further improve the effluent quality before it is discharged to the receiving environment (sea, river, lake, wet lands, ground, etc.). More than one tertiary treatment process may be used at any treatment plant. If disinfection is practised, it is always the final process. It is also called "effluent polishing".

Filtration

Sand filtration removes much of the residual suspended matter. Filtration over activated carbon, also called *carbon adsorption,* removes residual toxins.

Lagoons or Ponds

A sewage treatment plant and lagoon in Everett, Washington, USA.

Lagoons or ponds provide settlement and further biological improvement through storage in large man-made ponds or lagoons. These lagoons are highly aerobic and colonization by native macrophytes, especially reeds, is often encouraged. Small filter-feeding invertebrates such as *Daphnia* and species of *Rotifera* greatly assist in treatment by removing fine particulates.

Biological Nutrient Removal

Biological nutrient removal (BNR) is regarded by some as a type of secondary treatment process, and by others as a tertiary (or "advanced") treatment process.

Wastewater may contain high levels of the nutrients nitrogen and phosphorus. Excessive release to the environment can lead to a buildup of nutrients, called eutrophication, which can in turn encourage the overgrowth of weeds, algae, and cyanobacteria (blue-green algae). This may cause an algal bloom, a rapid growth in the population of algae. The algae numbers are unsustainable and eventually most of them die. The decomposition of the algae by bacteria uses up so much of the oxygen in the water that most or all of the animals die, which creates more organic matter for the bacteria to decompose. In addition to causing deoxygenation, some algal species produce toxins that contaminate drinking water supplies. Different treatment processes are required to remove nitrogen and phosphorus.

Nitrogen Removal

Nitrogen is removed through the biological oxidation of nitrogen from ammonia to nitrate (nitrification), followed by denitrification, the reduction of nitrate to nitrogen gas. Nitrogen gas is released to the atmosphere and thus removed from the water.

Nitrification itself is a two-step aerobic process, each step facilitated by a different type of bacteria. The oxidation of ammonia $\left(NH_3\right)$ to nitrite $\left(NO_2^-\right)$ is most often facilitated by *Nitrosomonas* spp. ("nitroso" referring to the formation of a nitroso functional group). Nitrite oxidation to nitrate $\left(NO_3^-\right)$, though traditionally believed to be facilitated by *Nitrobacter* spp. (nitro referring the formation of a nitro functional group), is now known to be facilitated in the environment almost exclusively by *Nitrospira* spp.

Denitrification requires anoxic conditions to encourage the appropriate biological communities to form. It is facilitated by a wide diversity of bacteria. Sand filters, lagooning and reed beds can all be used to reduce nitrogen, but the activated sludge process (if designed well) can do the job the most easily. Since denitrification is the reduction of nitrate to dinitrogen (molecular nitrogen) gas, an electron donor is needed. This can be, depending on the waste water, organic matter (from feces), sulfide, or an added donor like methanol. The sludge in the anoxic tanks (denitrification tanks) must be mixed well (mixture of recirculated mixed liquor, return activated sludge [RAS], and raw influent) e.g. by using submersible mixers in order to achieve the desired denitrification.

Sometimes the conversion of toxic ammonia to nitrate alone is referred to as tertiary treatment.

Over time, different treatment configurations have evolved as denitrification has become more sophisticated. An initial scheme, the Ludzack–Ettinger Process, placed an anoxic treatment zone before the aeration tank and clarifier, using the return activated sludge (RAS) from the clarifier as a nitrate source. Influent wastewater (either raw or as effluent from primary clarification) serves as the electron source for the facultative bacteria to metabolize carbon, using the inorganic nitrate as a source of oxygen instead of dissolved molecular oxygen. This denitrification scheme was naturally limited to the amount of soluble nitrate present in the RAS. Nitrate reduction was limited because RAS rate is limited by the performance of the clarifier.

The "Modified Ludzak–Ettinger Process" (MLE) is an improvement on the original concept, for it recycles mixed liquor from the discharge end of the aeration tank to the head of the anoxic tank to provide a consistent source of soluble nitrate for the facultative bacteria. In this instance, raw wastewater continues to provide the electron source, and sub-surface mixing maintains the bacteria in contact with both electron source and soluble nitrate in the absence of dissolved oxygen.

Many sewage treatment plants use centrifugal pumps to transfer the nitrified mixed liquor from the aeration zone to the anoxic zone for denitrification. These pumps are often referred to as *Internal*

Mixed Liquor Recycle (IMLR) pumps. IMLR may be 200% to 400% the flow rate of influent wastewater (Q). This is in addition to Return Activated Sludge (RAS) from secondary clarifiers, which may be 100% of Q. (Therefore, the hydraulic capacity of the tanks in such a system should handle at least 400% of annual average design flow (AADF). At times, the raw or primary effluent wastewater must be carbon-supplemented by the addition of methanol, acetate, or simple food waste (molasses, whey, plant starch) to improve the treatment efficiency. These carbon additions should be accounted for in the design of a treatment facility's organic loading. Further modifications to the MLE were to come: Bardenpho and Biodenipho processes include additional anoxic and oxidative processes to further polish the conversion of nitrate ion to molecular nitrogen gas. Use of an anaerobic tank following the initial anoxic process allows for luxury uptake of phosphorus by bacteria, thereby biologically reducing orthophosphate ion in the treated wastewater. Even newer improvements, such as Anammox Process, interrupt the formation of nitrate at the nitrite stage of nitrification, shunting nitrite-rich mixed liquor activated sludge to treatment where nitrite is then converted to molecular nitrogen gas, saving energy, alkalinity, and secondary carbon sourcing. Anammox (ANaerobic AMMonia OXidation) works by artificially extending detention time and preserving denitrifying bacteria through the use of substrate added to the mixed liquor and continuously recycled from it prior to secondary clarification. Many other proprietary schemes are being deployed, including DEMON, Sharon-ANAMMOX, ANITA-Mox, and DeAmmon. The bacteria Brocadia anammoxidans can remove ammonium from waste water through anaerobic oxidation of ammonium to hydrazine, a form of rocket fuel.

Phosphorus Removal

Every adult human excretes between 200 and 1,000 grams (7.1 and 35.3 oz) of phosphorus annually. Studies of United States sewage in the late 1960s estimated mean per capita contributions of 500 grams (18 oz) in urine and feces, 1,000 grams (35 oz) in synthetic detergents, and lesser variable amounts used as corrosion and scale control chemicals in water supplies. Source control via alternative detergent formulations has subsequently reduced the largest contribution, but the content of urine and feces will remain unchanged. Phosphorus removal is important as it is a limiting nutrient for algae growth in many fresh water systems. (For a description of the negative effects of algae, *see* Nutrient removal). It is also particularly important for water reuse systems where high phosphorus concentrations may lead to fouling of downstream equipment such as reverse osmosis.

Phosphorus can be removed biologically in a process called enhanced biological phosphorus removal. In this process, specific bacteria, called polyphosphate-accumulating organisms (PAOs), are selectively enriched and accumulate large quantities of phosphorus within their cells (up to 20 percent of their mass). When the biomass enriched in these bacteria is separated from the treated water, these biosolids have a high fertilizer value.

Phosphorus removal can also be achieved by chemical precipitation, usually with salts of iron (e.g. ferric chloride), aluminum (e.g. alum), or lime. This may lead to excessive sludge production as hydroxides precipitate and the added chemicals can be expensive. Chemical phosphorus removal requires significantly smaller equipment footprint than biological removal, is easier to operate and is often more reliable than biological phosphorus removal. Another method for phosphorus removal is to use granular laterite.

Some systems use both biological phosphorus removal and chemical phosphorus removal. The chemical phosphorus removal in those systems may be used as a backup system, for use when the biological phosphorus removal is not removing enough phosphorus, or may be used continuously. In either case, using both biological and chemical phosphorus removal has the advantage of not increasing sludge production as much as chemical phosphorus removal on its own, with the disadvantage of the increased initial cost associated with installing two different systems.

Once removed, phosphorus, in the form of a phosphate-rich sewage sludge, may be dumped in a landfill or used as fertilizer. In the latter case, the treated sewage sludge is also sometimes referred to as biosolids.

Disinfection

The purpose of disinfection in the treatment of waste water is to substantially reduce the number of microorganisms in the water to be discharged back into the environment for the later use of drinking, bathing, irrigation, etc. The effectiveness of disinfection depends on the quality of the water being treated (e.g., cloudiness, pH, etc.), the type of disinfection being used, the disinfectant dosage (concentration and time), and other environmental variables. Cloudy water will be treated less successfully, since solid matter can shield organisms, especially from ultraviolet light or if contact times are low. Generally, short contact times, low doses and high flows all militate against effective disinfection. Common methods of disinfection include ozone, chlorine, ultraviolet light, or sodium hypochlorite. Monochloramine, which is used for drinking water, is not used in the treatment of waste water because of its persistence. After multiple steps of disinfection, the treated water is ready to be released back into the water cycle by means of the nearest body of water or agriculture. Afterwards, the water can be transferred to reserves for everyday human uses.

Chlorination remains the most common form of waste water disinfection in North America due to its low cost and long-term history of effectiveness. One disadvantage is that chlorination of residual organic material can generate chlorinated-organic compounds that may be carcinogenic or harmful to the environment. Residual chlorine or chloramines may also be capable of chlorinating organic material in the natural aquatic environment. Further, because residual chlorine is toxic to aquatic species, the treated effluent must also be chemically dechlorinated, adding to the complexity and cost of treatment.

Ultraviolet (UV) light can be used instead of chlorine, iodine, or other chemicals. Because no chemicals are used, the treated water has no adverse effect on organisms that later consume it, as may be the case with other methods. UV radiation causes damage to the genetic structure of bacteria, viruses, and other pathogens, making them incapable of reproduction. The key disadvantages of UV disinfection are the need for frequent lamp maintenance and replacement and the need for a highly treated effluent to ensure that the target microorganisms are not shielded from the UV radiation (i.e., any solids present in the treated effluent may protect microorganisms from the UV light). In the United Kingdom, UV light is becoming the most common means of disinfection because of the concerns about the impacts of chlorine in chlorinating residual organics in the wastewater and in chlorinating organics in the receiving water. Some sewage treatment systems in Canada and the US also use UV light for their effluent water disinfection.

Ozone (O_3) is generated by passing oxygen (O_2) through a high voltage potential resulting in a third oxygen atom becoming attached and forming O_3. Ozone is very unstable and reactive and oxidizes most organic material it comes in contact with, thereby destroying many pathogenic microorganisms. Ozone is considered to be safer than chlorine because, unlike chlorine which has to be stored on site (highly poisonous in the event of an accidental release), ozone is generated on-site as needed from the oxygen in the ambient air. Ozonation also produces fewer disinfection by-products than chlorination. A disadvantage of ozone disinfection is the high cost of the ozone generation equipment and the requirements for special operators.

Fourth Treatment Stage

Micropollutants such as pharmaceuticals, ingredients of household chemicals, chemicals used in small businesses or industries, environmental persistent pharmaceutical pollutant (EPPP) or pesticides may not be eliminated in the conventional treatment process (primary, secondary and tertiary treatment) and therefore lead to water pollution. Although concentrations of those substances and their decomposition products are quite low, there is still a chance of harming aquatic organisms. For pharmaceuticals, the following substances have been identified as "toxicologically relevant": substances with endocrine disrupting effects, genotoxic substances and substances that enhance the development of bacterial resistances. They mainly belong to the group of environmental persistent pharmaceutical pollutants. Techniques for elimination of micropollutants via a fourth treatment stage during sewage treatment are implemented in Germany, Switzerland, Sweden and the Netherlands and tests are ongoing in several other countries. Such process steps mainly consist of activated carbon filters that adsorb the micropollutants. The combination of advanced oxidation with ozone followed by GAC, Granulated Activated Carbon, has been suggested as a cost-effective treatment combination for pharmaceutical residues. Also the use of enzymes such as the enzyme laccase is under investigation. A new concept which could provide an energy-efficient treatment of micropollutants could be the use of laccase secreting fungi cultivated at a wastewater treatment plant to degrade micropollutants and at the same time to provide enzymes at a cathode of a microbial biofuel cells. Microbial biofuel cells are investigated for their property to treat organic matter in wastewater.

To reduce pharmaceuticals in water bodies, also "source control" measures are under investigation, such as innovations in drug development or more responsible handling of drugs.

Odor Control

Odors emitted by sewage treatment are typically an indication of an anaerobic or "septic" condition. Early stages of processing will tend to produce foul-smelling gases, with hydrogen sulfide being most common in generating complaints. Large process plants in urban areas will often treat the odors with carbon reactors, a contact media with bio-slimes, small doses of chlorine, or circulating fluids to biologically capture and metabolize the noxious gases. Other methods of odor control exist, including addition of iron salts, hydrogen peroxide, calcium nitrate, etc. to manage hydrogen sulfide levels.

High-density solids pumps are suitable for reducing odors by conveying sludge through hermetic closed pipework.

Energy Requirements

For conventional sewage treatment plants, around 30 percent of the annual operating costs is usually required for energy. The energy requirements vary with type of treatment process as well as wastewater load. For example, constructed wetlands have a lower energy requirement than activated sludge plants, as less energy is required for the aeration step. Sewage treatment plants that produce biogas in their sewage sludge treatment process with anaerobic digestion can produce enough energy to meet most of the energy needs of the sewage treatment plant itself.

In conventional secondary treatment processes, most of the electricity is used for aeration, pumping systems and equipment for the dewatering and drying of sewage sludge. Advanced wastewater treatment plants, e.g. for nutrient removal, require more energy than plants that only achieve primary or secondary treatment.

Sludge Treatment and Disposal

The sludges accumulated in a wastewater treatment process must be treated and disposed of in a safe and effective manner. The purpose of digestion is to reduce the amount of organic matter and the number of disease-causing microorganisms present in the solids. The most common treatment options include anaerobic digestion, aerobic digestion, and composting. Incineration is also used, albeit to a much lesser degree. The use of a green approach, such as phytoremediation, has been recently proposed as a valuable tool to improve sewage sludge contaminated by trace elements and POPs.

Sludge treatment depends on the amount of solids generated and other site-specific conditions. Composting is most often applied to small-scale plants with aerobic digestion for mid-sized operations, and anaerobic digestion for the larger-scale operations.

The sludge is sometimes passed through a so-called pre-thickener which de-waters the sludge. Types of pre-thickeners include centrifugal sludge thickenersrotary drum sludge thickeners and belt filter presses. Dewatered sludge may be incinerated or transported offsite for disposal in a landfill or use as an agricultural soil amendment.

Environment Aspects

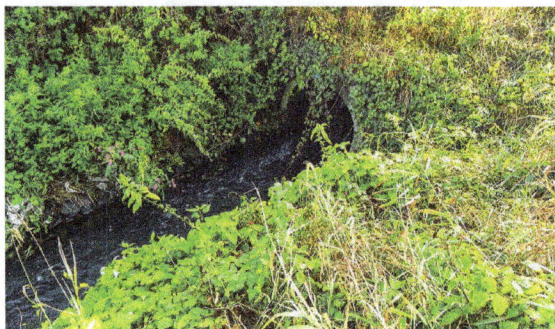

Treated water from WWTP Děčín, Czech Republic.

Treated water drained to the Elbe river, Děčín, Czech Republic.

The outlet of the Karlsruhe sewage treatment plant flows into the Alb.

Many processes in a wastewater treatment plant are designed to mimic the natural treatment processes that occur in the environment, whether that environment is a natural water body or the ground. If not overloaded, bacteria in the environment will consume organic contaminants, although this will reduce the levels of oxygen in the water and may significantly change the overall ecology of the receiving water. Native bacterial populations feed on the organic contaminants, and the numbers of disease-causing microorganisms are reduced by natural environmental conditions such as predation or exposure to ultraviolet radiation. Consequently, in cases where the receiving environment provides a high level of dilution, a high degree of wastewater treatment may not be required. However, recent evidence has demonstrated that very low levels of specific contaminants in wastewater, including hormones (from animal husbandry and residue from human hormonal contraception methods) and synthetic materials such as phthalates that mimic hormones in their action, can have an unpredictable adverse impact on the natural biota and potentially on humans if the water is re-used for drinking water. In the US and EU, uncontrolled discharges of wastewater to the environment are not permitted under law, and strict water quality requirements are to be met, as clean drinking water is essential. A significant threat in the coming decades will be the increasing uncontrolled discharges of wastewater within rapidly developing countries.

Effects on Biology

Sewage treatment plants can have multiple effects on nutrient levels in the water that the treated sewage flows into. These nutrients can have large effects on the biological life in the water in contact with the effluent. Stabilization ponds (or sewage treatment ponds) can include any of the following:

- Oxidation ponds, which are aerobic bodies of water usually 1–2 metres (3 ft 3 in–6 ft 7 in) in depth that receive effluent from sedimentation tanks or other forms of primary treatment:

  ◦ Dominated by algae.

- Polishing ponds are similar to oxidation ponds but receive effluent from an oxidation pond or from a plant with an extended mechanical treatment.

  ◦ Dominated by zooplankton.

- Facultative lagoons, raw sewage lagoons, or sewage lagoons are ponds where sewage is added with no primary treatment other than coarse screening. These ponds provide effective treatment when the surface remains aerobic; although anaerobic conditions may develop near the layer of settled sludge on the bottom of the pond.

- Anaerobic lagoons are heavily loaded ponds.

  ○ Dominated by bacteria.

- Sludge lagoons are aerobic ponds, usually 2 to 5 metres (6 ft 7 in to 16 ft 5 in) in depth, that receive anaerobically digested primary sludge, or activated secondary sludge under water.

  ○ Upper layers are dominated by algae.

Phosphorus limitation is a possible result from sewage treatment and results in flagellate-dominated plankton, particularly in summer and fall.

A phytoplankton study found high nutrient concentrations linked to sewage effluents. High nutrient concentration leads to high chlorophyll a concentrations, which is a proxy for primary production in marine environments. High primary production means high phytoplankton populations and most likely high zooplankton populations, because zooplankton feed on phytoplankton. However, effluent released into marine systems also leads to greater population instability.

The planktonic trends of high populations close to input of treated sewage is contrasted by the bacterial trend. In a study of *Aeromonas* spp. in increasing distance from a wastewater source, greater change in seasonal cycles was found the furthest from the effluent. This trend is so strong that the furthest location studied actually had an inversion of the *Aeromonas* spp. cycle in comparison to that of fecal coliforms. Since there is a main pattern in the cycles that occurred simultaneously at all stations it indicates seasonal factors (temperature, solar radiation, phytoplankton) control of the bacterial population. The effluent dominant species changes from *Aeromonas caviae* in winter to *Aeromonas sobria* in the spring and fall while the inflow dominant species is *Aeromonas caviae*, which is constant throughout the seasons.

Reuse

With suitable technology, it is possible to reuse sewage effluent for drinking water, although this is usually only done in places with limited water supplies, such as Windhoek and Singapore.

In arid countries, treated wastewater is often used in agriculture. For example, in Israel, about 50 percent of agricultural water use (total use was one billion cubic metres $\left(3.5 \times 10^{10} \text{ cu ft}\right)$ in 2008) is provided through reclaimed sewer water. Future plans call for increased use of treated sewer water as well as more desalination plants as part of water supply and sanitation in Israel.

Constructed wetlands fed by wastewater provide both treatment and habitats for flora and fauna. Another example for reuse combined with treatment of sewage are the East Kolkata Wetlands in India. These wetlands are used to treat Kolkata's sewage, and the nutrients contained in the wastewater sustain fish farms and agriculture.

References

- Frerichs, Ralph R. "History of the Chelsea Waterworks". John Snow. Fielding School of Public Health, University of California, Los Angeles. Retrieved 2016-07-09

- Treating-water-pollution: water-pollution.org.uk, Retrieved 1 February, 2019

- Hargrove, Maddy; Hargrove, Mic (2006). Freshwater Aquariums for Dummies. Hoboken, New Jersey: Wiley Publishing, Inc. P. 181. ISBN 9780470051030

- Swistock, Bryan. "Methane Gas and Its Removal from Wells in Pennsylvania". PSU. Retrieved 18 June 2014

- Van Trump, James Ian; Coates, John D. (2008-12-18). "Thermodynamic targeting of microbial perchlorate reduction by selective electron donors". The ISME Journal. 3 (4): 466–476. Doi:10.1038/ismej.2008.119. PMID 19092865

- Article on "Water treatment solution: Filtration", retrieved on 15 October 2013 from http://www.lenntech.com/chemistry/filtration.htm

Permissions

Index

www.ingramcontent.com/pod-product-compliance
Lightning Source LLC
Chambersburg PA
CBHW061311190326
41458CB00011B/3776